Science and Fict

Science and Fiction – A Springer Series

This collection of entertaining and thought-provoking books will appeal equally to science buffs, scientists and science-fiction fans. It was born out of the recognition that scientific discovery and the creation of plausible fictional scenarios are often two sides of the same coin. Each relies on an understanding of the way the world works, coupled with the imaginative ability to invent new or alternative explanations—and even other worlds. Authored by practicing scientists as well as writers of hard science fiction, these books explore and exploit the borderlands between accepted science and its fictional counterpart. Uncovering mutual influences, promoting fruitful interaction, narrating and analyzing fictional scenarios, together they serve as a reaction vessel for inspired new ideas in science, technology, and beyond.

Whether fiction, fact, or forever undecidable: the Springer Series "Science and Fiction" intends to go where no one has gone before!

Its largely non-technical books take several different approaches. Journey with their authors as they

- Indulge in science speculation – describing intriguing, plausible yet unproven ideas;
- Exploit science fiction for educational purposes and as a means of promoting critical thinking;
- Explore the interplay of science and science fiction – throughout the history of the genre and looking ahead;
- Delve into related topics including, but not limited to: science as a creative process, the limits of science, interplay of literature and knowledge;
- Tell fictional short stories built around well-defined scientific ideas, with a supplement summarizing the science underlying the plot.

Readers can look forward to a broad range of topics, as intriguing as they are important. Here just a few by way of illustration:

- Time travel, superluminal travel, wormholes, teleportation
- Extraterrestrial intelligence and alien civilizations
- Artificial intelligence, planetary brains, the universe as a computer, simulated worlds
- Non-anthropocentric viewpoints
- Synthetic biology, genetic engineering, developing nanotechnologies
- Eco/infrastructure/meteorite-impact disaster scenarios
- Future scenarios, transhumanism, posthumanism, intelligence explosion
- Virtual worlds, cyberspace dramas
- Consciousness and mind manipulation

More information about this series at
https://link.springer.com/bookseries/11657

Eric Choi

Just Like Being There

A Collection of Science Fiction Short Stories

 Springer

Eric Choi
Toronto, ON, Canada

ISSN 2197-1188 ISSN 2197-1196 (electronic)
Science and Fiction
ISBN 978-3-030-91604-6 ISBN 978-3-030-91605-3 (eBook)
https://doi.org/10.1007/978-3-030-91605-3

Abstract Theme of Spaceman and Planet Earth, credit: Sergey Nivens/Shutterstock.com

This Springer imprint is published by the registered company Springer Nature Switzerland AG
The registered company address is: Gewerbestrasse 11, 6330 Cham, Switzerland

"The fifteen stories comprising Eric Choi's *Just Like Being There* take readers on a fun and fictional romp through space and time, from alternate histories to daring future missions. Some of the narratives are sobering revisits of past events on Earth while others launch us on adventures in deep space. All are imaginative and captivating."

—Robert Thirsk, *Former CSA Astronaut*

"Eric Choi's story collection takes us on an epic literary road trip across the red sands of Mars, down the mysterious tunnels of an asteroid, into our ocean's inky depths, and even plunges into the remote crevasses of our brain. His narratives are a perfect fusion of meticulous scientific research and imagination, a classic hard SF approach that ponders territories just beyond our reach. While Choi's alternative histories and fictional tales cover a wide range of topics, each one is an invitation to see the world anew. He challenges us to let go of stereotypes and assumptions, and he does so with great subtlety and charm. The afterwords following each story provide fascinating background information that contextualizes the narratives and provide intriguing insights into politics, science, and history. Clear your agenda. This book is impossible to put down."

—Bettina Forget, *Director of the Artist-in-Residence Program at the SETI Institute*

"Eric Choi has created an intellectually engaging collection of stories with afterwords discussing the meticulously researched science of each. A real tour de force."

—Michael Brotherton, *Professor of Astronomy at the University of Wyoming,*
Editor of Science Fiction by Scientists

"When an award-winning writer and editor is also an aerospace engineer you get some of the best hard SF stories around. In *Just Like Being There*, Eric Choi writes with clarity, grace, and imagination as he leads us through an astonishing breadth of topics and themes from near-future stories of astronauts doing dangerous work to alternate history tales on Captain Nemo. The fifteen stories in this book are entertaining, exciting, and thoughtful by turn. This collection deserves a huge audience. Very highly recommended."

—Rick Wilber, *Author and Visiting Professor of Genre Fiction at Western Colorado University, Administrator of the Dell Magazines Award for Undergraduate Excellence in Science Fiction and Fantasy Writing*

"Enjoyable and thought provoking. This collection of short stories is a must read for those who have an interest in aviation, space, and scientific innovation. Choi's blend of futuristic settings and historical what-ifs challenged a number of my assumptions about the science fiction genre and made me look at our history–and some of the treasures in our collection–in a completely different light."

—Christopher Kitzan, *Director General of the Canada Aviation and Space Museum*

Preface

My dual career as a writer and aerospace engineer was undoubtedly influenced by two childhood memories. One of my earliest recollections is of making a LEGO model of the Voyager spacecraft, cutting pictures of Saturn out of a magazine and imagining that my little plastic space probe had taken them. Another was a grainy television image of an unfortunate man in a red shirt being killed by an acid-secreting silicon-based lifeform in what I later learned was the original series *Star Trek* episode "The Devil in the Dark". For as long as I can remember, science and science fiction have been like two sides of a coin.

Star Trek might have remained the extent of my science fiction universe (not that there is anything wrong with that) had it not been for a childhood friend named Leslie Gelberger who introduced me to a story called "The Road Not Taken" by Eric G. Iverson in the November 1985 issue of *Analog* magazine. I was thrilled to think there was apparently a science fiction writer out there also named Eric (alas, the table of contents outed him as Harry Turtledove). The ending of "The Road Not Taken" lingered with me for days, and it fostered a dream that perhaps one day I might also write science fiction stories that are equally thought-provoking.

My fascination with space exploration began in childhood and has never diminished. The only question was whether I would pursue a career in engineering or the sciences. My decision was made in high school after reading a *National Geographic* article about Voyager which described the miracles performed by the engineers that enabled the spacecraft to continue its mission beyond the original destinations of Jupiter and Saturn. Designed and launched in the 1970s, Voyager used a mechanical tape recorder to store data. If the tape recorder was turned on, then a thruster would have to be fired to keep the spacecraft steady during the long exposures needed for taking images in the dark outer Solar System. My teenage self was amazed by the very notion that there are people whose job is to do such astounding things.

So, I decided to become an engineer, enrolling in the engineering science programme at the University of Toronto. During the third year of my studies when I began a specialization in aerospace, my friend Raakesh Persaud made me aware of the Isaac Asimov Award for Undergraduate Excellence in Science Fiction and Fantasy Writing. This is a contest established by *Asimov's Science Fiction* magazine and the International Association of the Fantastic in the Arts whose goal is to encourage emerging student writers of speculative fiction. I entered the contest, and all these years later it is still difficult to describe my shock at becoming the first winner of the Asimov Award (now called the Dell Magazines Award) in the first year it was offered. My winning story was called "Dedication", which is appropriately the opening story of this collection.

Following my undergraduate degree, I completed graduate studies at the University of Toronto Institute for Aerospace Studies with a specialization in satellite orbit and attitude dynamics (no doubt a reflection of my early fascination with the spacecraft pointing requirements of the Voyager mission). Over the course of my aerospace engineering career, I have had the privilege of working on a number of space projects including the Quantum Encryption and Science Satellite (QEYSSat), the Meteorology (MET) instruments on the Phoenix Mars Lander, the Canadarm2 manipulator on the International Space Station, the RADARSAT-1 Earth observation satellite, and the Measurements of Pollution in the Troposphere (MOPITT) instrument on the Terra satellite. In 2009, I was one of the top 40 finalists (out of 5,351 applicants) in the Canadian Space Agency's astronaut recruitment campaign.

I am not a prolific writer. Since winning the Dell Award as a student all those years ago, I have only written nineteen stories that have appeared in twenty-nine publications. Fourteen of my science and engineering themed stories are reprinted here, plus a new story called "A Sky and a Heaven" which closes this collection. Readers should be aware that the reprint stories may not be identical to the versions that were published elsewhere. This is because I went back to the manuscripts as I had originally written them and in some cases made minor revisions. For example, in the opening story "Dedication" I changed what I had called a "palmtop" computer to the modern term of a tablet. The high school mentioned in the title story "Just Like Being There" was renamed for Leslie Gelberger, the childhood friend who introduced me to science fiction short stories. I moved out the dates in "From a Stone" because as of the publication of this collection humans have not yet resumed crewed voyages beyond low Earth orbit. In general, however, I was pleasantly surprised at how well my stories have held up over time. For each story, I have written a new afterword that discusses the engineering and science behind them.

The short story is a beautiful literary form because each paragraph or sentence or even word can have so much more impact than they would in a longer work like a novel. It is also an ideal format for exploring new ideas. To quote the American editor and publisher Patrick Nielsen Hayden: "Short fiction is…the R&D laboratory in which SF [science fiction] constantly reinvents itself. Novels are all very well…but short fiction is where SF writers take chances, stir the pot, kick out the jams. It's no accident that short fiction has been crucial to each period of major ferment and invention in the modern history of the field. It is, if you will, the garage rock of SF".[1]

The title story "Just Like Being There" was first published in a volume of a series called *Tales From the Wonder Zone* that was created by the biologist and writer Julie Czerneda which uses science fiction to illustrate scientific concepts for elementary school students. One of the greatest challenges of our time is the lack of scientific literacy amongst both the general public and the political leadership of many countries. In the era of pandemics and climate change, not having an inherent understanding of fundamental principles like exponential growth, statistical significance, the sensitivity of models to initial and boundary conditions, and the risks of extrapolating data beyond established domains of validity is putting all of us in peril.

Science fiction is the literature of change, the genre that examines the implications—both beneficial and dangerous—of new sciences and technologies. It is uniquely able to do this because science fiction is not just about what is and what was, but what could be. This is what appeals most to me about the genre. A compelling science fiction story can take a reader to any point in space and time, from the farthest reaches of the Universe to the depths of the human soul. It really can be just like being there.

Toronto, Canada   Eric Choi
February 2022

---

[1] Hayden, Patrick Nielsen (editor). *Starlight 1*, Tor Books, 1996, Pg. 9.

Contents

Dedication

There is no gold on the Plains of Gold.

The region is more commonly known by its Latin name, Chryse Planitia. Not a trace of gold was in evidence. The predominant colour, as it was everywhere on Mars, was red. Red in every shade, complexion, and hue imaginable.

Sharp, angular pieces of rusty limonite were scattered about. The rocks ranged in size from a few centimetres to a two-metre monster nicknamed Big Joe. The entire vista was blanketed overhead by a deceptively warm-looking sky of salmon pink.

None of this mattered to Oleg Solovyov. He was not here to gawk at the scenery. As the engineering specialist for the Ares 4 mission, the examination of the human relic before him was his unique responsibility. Sightseeing could, and would, have to wait.

Had Solovyov a bit of imagination, he might have described the jeep-sized craft as bearing a strange resemblance to a robotic camel. A parabolic dish antenna atop a narrow boom could be thought of as the creature's head and neck. For humps, two cylindrical silos housing the cameras sprouted from its top, and in lieu of a tail the machine possessed a stubby mechanical arm for taking soil samples.

But a most unusual looking camel, at that. The "head" was stuck in the middle of and behind the "humps", while the "tail" jutted out haphazardly in front. Besides, the craft only had three legs.

Through his spacesuit radio, he could hear an enthusiastic monologue delivered by Dr. Michèle Lafrenière, the Ares 4 science specialist. She had

E. Choi, *Just Like Being There*, Science and Fiction,
https://doi.org/10.1007/978-3-030-91605-3_1

been taking samples from the regolith—the surface layer of dust and rock—at a site ninety metres behind Solovyov, but she was now in the process of recording a video of her work for later TV and Internet distribution on Earth. She went on to talk about the subsurface permafrost layer that she had also taken a core sample of, and her hopes that the presence of water meant that Mars once held or still harboured some form of life.

Solovyov shook his head. Lafrenière had been an adjunct professor of planetary science at the International Space University in Strasbourg before being accepted into Project Ares. He completely failed to understand how a person of her intelligence could actually enjoy wasting time on such meaningless activities. He, at least, had other things to worry about.

Things like the metal pipe he was trying to cut off that refused to give. For the past five minutes, he had been trying to get a sample of one of the old spacecraft's nitrogen lines without success. Cutting metal in the Martian environment was a stranger experience than Solovyov had expected. The biggest difference from doing so on Earth was the absence of any sound or sparks from the friction of his blade, robbing him of the familiar audio and visual cues.

Lafrenière in the meantime had completed her narration, and Solovyov's helmet speakers fell silent. That meant he was next. He grunted, and pressed the tool harder against the metal. Slowly he could feel the circular blade working its way until with a sudden jerk the cutter went through, and the metal sample snapped off easily into Solovyov's grip.

"Done?" asked a new voice on the radio.

Solovyov turned around and presently another spacesuited figure came into view. Shaun Christopher, the mission commander, stood before him holding a small camera in his right hand. Having finished recording Lafrenière's presentation he was now, Solovyov knew, about to pester him.

"Yes," Solovyov replied curtly. He gestured to the camera. "Is that truly necessary?"

Christopher hefted the device in his hand. "Well, Michèle just completed a short talk on her work, and Doc gave a little tour of the rover earlier, so I guess it's your turn, Ollie."

Solovyov winced. He despised that anglicized nickname Christopher had bestowed upon him, but kept his mouth shut in the interest of crew harmony.

"Commander Christopher–"

"Shaun."

"—there is so much work to be done. Perhaps the time could be better spent..."

"Look, Ollie, this doesn't take any time at all, and it's important." Christopher sighed audibly. "I don't get it. Why are you always like this?"

"I do not believe it is important," Solovyov said stubbornly.

Christopher spread out his arms in a pleading gesture. "Well, it is. You know, 'No bucks, no Buck Rogers.'" He paused. "Buck Rogers?"

"What is a Buck Rogers?"

"Just talk about the lander, talk about your work, OK? Please? Everyone else's done their song and dance. It will look strange if you're left out."

"Very well. But briefly." Solovyov gritted his teeth.

"Thank you." Christopher raised the camera until its eyepiece was against the tinted faceplate of his helmet. His left hand made a few adjustments to the lens before he started recording. "Here's Oleg Solovyov, our engineering specialist. What is this old relic, and what kind of work have you been doing with it?"

"The old spacecraft you see before you is the Viking 1 Lander." Solovyov took a step out of the way and gestured at it with his right hand. "It was launched in the last century, and was the first spacecraft to successfully soft-land on Mars. The Viking is of great engineering importance. From it we can learn about the long-term effects of the Martian environment on materials and machines. I have collected some parts of the craft for analysis on Earth." Specifically, he had taken a piece of tubing, a length of cable, the collector head from the Viking's sampler arm and one of the lander's two cameras. "This information will be vital if we build a permanent base on Mars."

"I did a video survey of the Viking before you started taking samples, to get a record of its original condition." Christopher said. "Besides its reddish tint, the lander shows remarkably few signs of wear."

"Yes, physically the Viking is in exceptional condition. Except for the dust." He turned from the camera momentarily to look at the lander. Its entire metal frame was coated with a fine rust-like layer, and each of its three legs was embraced by a fillet of the material. Such was its mute testimony to the raging dust storms that often engulfed the planet, darkening the sky and sending temperatures plummeting. The study of this fearsome phenomenon by Viking and other spacecraft had led terrestrial scientists to the theory of nuclear winter.

"Okay, that's fine," Christopher said as he lowered the camera. Solovyov let out a long breath. "You know, you really did very well."

Solovyov brightened at the unexpected compliment. "Thank you."

"Now, was that so bad?" the commander added.

Solovyov did not reply.

A moment later, the two astronauts were joined by a third. Having completed her geological sample collection tasks, Michèle Lafrenière approached her colleagues. Her right hand held a wire-frame mesh bag full of sample containers, while her left held the long corer device and various other tools.

Solovyov looked up as she neared, and Christopher followed his gaze. "I know *you're* done."

"Of course."

"All right, then. I guess it's time." The commander took a few steps up to Solovyov and handed him the camera. "Will you do the honours?"

He took the device from Christopher.

The scientific and engineering objectives of the EVA had been completed. But there was one more, very special task to perform.

Christopher walked toward the north-western face of the Viking—the side opposite from the robotic arm where Solovyov had been working. With Lafrenière at his side, the Russian followed their commander around the old spacecraft.

"How's your foot?" she asked.

Solovyov looked down at his right foot. He had tripped on a rock and fallen earlier in the afternoon, and the leg of his spacesuit still bore a ruddy stain from the mishap. "It is a little sore, but it is nothing."

"Clumsy," Lafrenière chided playfully.

Christopher positioned himself in front of the Viking. From the bottom of the camera, Solovyov released the tripod and extended the legs. Even though the camera was small and light, the appendages looked impossibly too thin to hold up its weight. Each of the three silvery telescoping segments resembled a segment of the rabbit-ear antennae of early TV sets. But Martian gravity was only two-fifths that of Earth, so the flimsy looking stand was more than adequate for the job.

Solovyov set the camera on the ground, made sure it was level, then looked into the eyepiece. He saw a flashing yellow light. "There is not much more room on this card," he reported.

"How much?" Christopher asked.

"16 minutes," Solovyov replied.

"Aw, that's fine. Plenty." Christopher stomped his feet and adjusted his position, then said, "Okay. Whenever you're ready."

Solovyov pressed some buttons on a keypad on top of the camera, adjusted the lens, looked into the eyepiece, and gestured for Christopher to start.

"This is Commander Christopher, speaking to you again from the surface of Mars. We're just about finished our work here at the western slope of Chryse Planitia, but before we leave, there's one more thing we have to do.

"You know, the reason I'm able to speak to you from Mars today is because of all the hard work people around the world put into making our voyage here possible. Not just the work done in the last decade or so with the Ares program, but the efforts of visionary men and women stretching all the way back to the last century.

"Today, I'd like to take this time to honour just one of those individuals. His name was Thomas Mutch. He was a scientist—a member of the team that sent this probe, the Viking 1 Lander," he gestured at the spacecraft, "here in 1976. Dr. Mutch saw planetary exploration as a grand adventure of the human spirit, and it was always his dream that one day people would walk the surface of Mars, maybe even..." Christopher reached out and laid a hand on the lander. "Maybe even pay his beloved Viking a visit.

"After his death in 1980, the Viking 1 Lander was renamed the Thomas A. Mutch Memorial Station in his honour." At this point, Christopher paused his monologue and unzipped a pouch on his left thigh. From the pouch, he produced a small, stainless steel object. "This plaque," he held it up to the camera, "was made then, in the hope that one day someone would bring it to Mars and attach it to the spacecraft. Well, it's taken many decades for the people of Earth to muster the will and the resources, but today it is the honour—and the privilege—of the Ares 4 crew to fulfil that dream."

With infinite care, Commander Shaun Christopher bolted the Mutch Station Plaque to the Viking 1 Lander.

When the task was completed, the commander took a few steps back and read the inscription aloud:

Dedicated to the memory of Tim Mutch, whose imagination, verve, and resolve contributed greatly to the exploration of the Solar System.

Finally, he took a deep breath and said, "I hereby dedicate the Thomas A. Mutch Memorial Station." With those words he brought himself smartly to attention, and saluted.

Beside Solovyov, Lafrenière put down her gear and saluted the monument as well—a gesture of unmitigated sentimentalism that the engineer wanted no part of.

By the time Solovyov returned to the rover with Christopher and Lafrenière, he was tired and sweaty. It was the same with every EVA. At first, he would look forward to the excursion as a chance to escape the cramped confines of the vehicle, but by the day's end he was always—well, tired and sweaty.

Solovyov vacuumed off the red dust that covered his spacesuit before stowing the suit, then waited for his turn at the personal hygiene facility. His mood began to improve after he had a chance to freshen up with moist towels and change into a standard powder-blue in-vehicle jumpsuit. Once he'd cleaned up, he went to the flight physician for an examination of his foot.

Dr. Wong Xuesen was the final member of what the media dubbed the Ares 4-some. He was a short, stocky man in his late thirties. He had bulging brown eyes and a straight-banged crop of black hair that—according to Christopher—gave him more than a passing resemblance to the Moe character in old *Three Stooges* programs. Solovyov had no idea what the commander was talking about.

"I think you strained a muscle," the physician said. He still spoke with a slight British accent, a remnant of his undergraduate days at Edinburgh University. The two men sat beside one another on a narrow bench in the rover's laboratory compartment, just behind the cab. Solovyov had rolled up his right pant leg and taken off his sock, while Dr. Wong gingerly examined his foot.

"Think?"

"You strained a muscle," Dr. Wong repeated as he opened his kit. He produced a vial and removed two pills from it. "Take these, and let me know if it still hurts tomorrow."

Solovyov took the pills and fingered them doubtfully.

"It's just ASA," Dr. Wong replied to the unspoken question.

Christopher was leaning against a bulkhead with his arms crossed, watching the exchange. The 49-year-old veteran astronaut was still as muscular and fit as a man a decade younger. Only the grey streaks in his once all-brown hair betrayed his age. Upon hearing Dr. Wong's prescription, he laughed. "Take two aspirin and call me in the morning! Some things never change. Any other recommendations, Doc?"

"Well," Dr. Wong began as he closed his kit, "I suppose he ought to go easy on that foot for a while. Give it a rest."

"Ah." Christopher raised an eyebrow and smiled deviously, as if he had heard exactly what he wanted to hear. "You know, Doc, of the four of us you're the only one who's never actually piloted the rover—outside of the simulator of course."

Although driving the rover was primarily the responsibility of Christopher and Solovyov, both Lafrenière and Wong had been trained in its operation in case of an emergency. Lafrenière had piloted the vehicle for several hours during their journey to the Viking site, but as Christopher had noted, Dr. Wong had no such practical experience.

While Solovyov rolled down his pant leg, Dr. Wong made a slow and deliberate show of stowing his kit. "But…but, surely, it's not terribly necessary that —"

Christopher held up his hands to silence him. "Sure, it's necessary. Doc, what if something happened to me, or Ollie…and Michèle, too. You could be the only one left to get us out of trouble."

"Now?" Dr. Wong whined.

"No better time." Christopher played his ace card. "Besides, it was your *medical opinion* that Ollie ought to give his foot a rest."

"What's this I hear about Dr. Wong driving?" Lafrenière asked as she entered from the rover's rear habitation compartment.

Christopher shrugged his shoulders, feigning innocence. "Aw, we just figured that it was about time Doc here got some practice driving the rover."

"Really?" Lafrenière surveyed the expression on Wong's face. "I think he looks scared."

"I am not scared!" Dr. Wong snapped.

"Look, Doc, just for a little while? At least an hour. Then I'll take over if you want." Christopher uncrossed his arms and took a few steps toward Dr. Wong. "It'll be nightfall in three hours. We'll have to stop then anyway."

"I agree with Commander Christopher," Solovyov chimed in.

An obviously exasperated Wong looked at Christopher, Solovyov, Lafrenière, and then back at Christopher. "All right, I'll do it," he sighed.

"Good!" Christopher said earnestly as he patted Dr. Wong on the shoulder.

Lafrenière ran her fingers through her curly red hair. "I guess I could go through the engine start checklist for him."

Christopher pointed a finger at the science specialist and smiled. "I like your attitude!"

He and Solovyov turned and made their way to the rover's forward cab, leaving Lafrenière and Dr. Wong in the laboratory compartment to stow their samples and gear before getting underway. Solovyov sat down at the engineering console at the rear-right of the cab, while Christopher went to the right-seat communications station in front of Solovyov to make his report to Mission Control in Darmstadt.

The Project Ares flights had been called "direct" missions. The small ships employed required no orbital assembly at a space station and the trajectory to Mars did not include a detour to Venus for a gravitational slingshot assist.

The Ares 4 mission had begun three years earlier with the launch of an uncrewed Earth Return Vehicle, or ERV, from Kennedy Space Centre. The ERV carried a payload of liquid hydrogen, a set of compressors, and a chemical processing unit. Upon arrival at Mars the ERV aerobraked into orbit, using the Martian atmosphere instead of fuel to slow down. It had landed on the north-western edge of Lunae Planum, near the southern rim of Kasei Vallis.

Once the Ares 4 base camp was established, the chemical processing plant used the liquid hydrogen cargo and the plentiful carbon dioxide of the Martian atmosphere to produce methane and water. These were, in turn, transformed into oxygen gas and liquid oxidizer, then back to hydrogen that was recycled to continue the process. Most of the propellant produced was used to fuel the ERV, but enough methane was left over to power the yet to arrive long-range rover.

Two years later, the Ares 4 Mars Excursion Vehicle, or MEV, was launched from Baikonur Cosmodrome carrying the four astronauts, provisions, and the pressurized rover currently occupied by Christopher and his crew. Engineers planned a simultaneous launch for the Ares 5 ERV that would accommodate another human expedition three years hence. Meanwhile, the Ares 4 MEV followed a radio beacon to the Lunae site, and landed next to the now fully refuelled ERV.

By manufacturing propellant on Mars instead of carrying it from Earth, the Ares direct plan dramatically reduced the launch mass of the spacecraft, and in direct proportion, reduced the cost of the missions enough to make them politically acceptable. But the scheme was not without its critics.

Many questioned the value of sending only four astronauts at a time, and worried about the tremendous physiological risks associated with the nominal sixteen month surface stays. Others criticized what they perceived to be a lack of backups and redundancy in some mission critical hardware, as well as the absence of continuous communication with Earth.

There were other problems as well. For one thing, the fragile daisy-chain of this tag-team approach was easily broken. The Ares 5 ERV, scheduled for launch a few days after Christopher and his crew were sent to Mars, had been delayed for almost two months due to mechanical glitches with its heavy-lift rocket. Still, science on a shoestring was deemed to be better than none at all, so the missions proceeded.

"…completed without incident. Status of consumables: Water 75.1%, fart juice 72.8%…" "Fart juice" was Christopher's nickname for the methane that fuelled the rover. "…oxygen 69.8%."

Solovyov saw Christopher frown as he read the figure from the environmental control status screen. "I believe our oh-two consumption during EVA is running a little higher than we expected. We'll check the status again at our next way point, and I'd like permission to cancel the next EVA at that time if Ollie and I deem it necessary." He paused and turned to Solovyov, who nodded. "Food status is OK…except Michèle ate the last shrimp salad…".

"Not guilty!" Lafrenière shouted playfully from the laboratory compartment.

Christopher grinned. "Well, that's all folks! EVA video and telemetry packet to follow. Talk to you again at the next way point. Christopher out." His fingers entered the appropriate commands on the touchscreen to transmit the message.

Wong and Lafrenière entered the cab just as Christopher finished his report. "Stop trying to give me a bad reputation," she lectured, wagging an accusing finger in Christopher's direction.

Christopher smiled and shrugged his shoulders. "But it's so much fun!"

Lafrenière muttered something Solovyov couldn't hear, and sat down at the science station beside him. Dr. Wong slid into the left-side driver's seat and strapped himself in, putting on a lap-belt and securing the shoulder harnesses.

As Solovyov and Christopher did the same, Lafrenière called up the engine start checklist on her console and began reading the items aloud. At each point, Dr. Wong either touched a control or brought up a screen and called back a confirmation. Christopher nodded with approval as the two followed the procedure to the letter.

By now, a low humming drone could be heard in the background. Solovyov could feel a very gentle vibration in his seat as the rover's methane engine came on-line. Having reviewed the last item on her checklist, Lafrenière turned off her screen.

"All done," she announced.

"Excellent, both of you," Christopher said. "Okay, whenever you're ready, Doc."

Dr. Wong hesitated for a moment before putting the rover in forward drive, releasing the brake, and gently stepping on the accelerator. The rover responded, timidly venturing forth at a conservative ten kilometres per hour.

While Christopher kept an eye on Dr. Wong, Solovyov, and Lafrenière settled back to take in the view provided by the cab's wide, wrap-around duraplex windshield—the only window on the rover. A boulder strewn gorge paralleled their path several metres to their right. The rocky terrain of Chryse Planitia slowly gave way to a dustier, almost lunar-like landscape, although many large stones were still in evidence. The landscape bore a striking resemblance to Death Valley, where the crew had spent some time training in basic geology.

Lafrenière unbuckled her seat restraints and stood. "I'm going to get a drink. Anybody want something?"

The three men replied in the negative, and Lafrenière exited the cab.

"You're doing fine, Doc," Christopher said.

"Uh, thank you." Dr. Wong muttered.

"See, it's not so bad," he added. Dr. Wong did not reply.

The commander snuck a peek at the door of the cab, then turned secretively to Solovyov and Dr. Wong. "Michèle is going kill me for telling you this, but it's so funny I have to –"

He was abruptly interrupted by a loud *bang*.

"What was that?" Dr. Wong gasped.

Solovyov furrowed his brow, trying to identify the noise. It sounded like a muffled howitzer had gone off outside, above and behind them.

"What's going on?" Lafrenière asked as she ran back into the cab.

Dr. Wong shot Christopher a questioning glance, but his foot was still on the accelerator.

Solovyov stared straight ahead out the window. Suddenly before his eyes the ground erupted in little puffs of dust.

Lafrenière returned to her seat and buckled her lap belt.

"Doc –" Christopher began, but he never finished the order.

The duraplex windshield right in front of Dr. Wong shattered into a bizarre, angular spider-web pattern as a meteorite struck the rover. Wong let out a strangled cry, and instinctively slammed the joystick to the right and floored the accelerator as if to avoid a non-existent obstacle. The rover skidded and began to fishtail.

"Turn, Doc!" Christopher shouted. "Turn into it!"

Dr. Wong pulled himself together enough to hear and understand Christopher's words. Manipulating the joystick purposefully now, he turned the rover into the direction of the skid, and slowly the vehicle began to straighten itself.

"The gorge!" Solovyov warned urgently. As Dr. Wong fought for control, the rover assumed a course that was taking it directly toward the menacing fissure.

Dr. Wong squinted his eyes and slammed on the brakes. The master alarm sounded and lights flashed on the control panel, indicating a failure of the rover's anti-skid system. All four wheels locked, and the astronauts sat helplessly as they slid into the gorge.

Solovyov was tossed about madly as the rover went down, his harness chafing at his jumpsuit as the straps held him in his seat. A horrendous scraping noise could be heard as jagged outcroppings of rock scoured the underside of the rover. Seconds later, the front of the vehicle hit the bottom of the ravine and crumpled, accompanied by a softer *thump* from inside the cab, as they came to a bone-jarring halt.

Christopher managed to reach a key to turn off the master alarm. The cab was silent, except for the sound of heavy breathing. Dr. Wong's left hand had the joystick in a virtual death grip, but his right hand was shaking. Beads of sweat covered his face and soaked the collar of his jumpsuit.

"Doc," Christopher turned to Wong, who was still trembling. "Doc! It's OK. Relax."

"I...I feel a little dizzy, I think..." the physician stammered.

"It's all right. Try to relax."

"OK..."

"Now, Doc, you're going to have to get us out of here." Wong opened his mouth to protest, but Christopher cut him off. "I can't get out of my seat. None of us can." The rover was pitched forward at a forty degree angle, pressing the crew against their harnesses. "It's all right. You can do it. Just stay calm."

Dr. Wong swallowed hard, and nodded silently.

"All right, carefully now." Christopher's voice was at once reassuring and stern. "Back us out, slow. Nice and slow."

Dr. Wong put the rover in reverse and gently pressed the accelerator. The engine revved up, and the rover backed out a few metres before the ground gave way and they slid back down.

"That's OK, we'll just try again. Do it once more. Gently."

Again the attempt failed, but on the third try the rear wheels found traction and with painful slowness the rover crawled up the side of the gorge. Gradually, the rocky bottom of the fissure receded, until what seemed like an eternity later they backed out over the rim and onto level ground.

Christopher let out a long breath and rubbed the bridge of his nose. "Good. Well done." He surveyed the cab. "Is everyone all right?"

"Dr. Lafrenière!" Solovyov exclaimed.

Michèle Lafrenière was unconscious, her head slumped forward in her seat. She had managed to buckle only her lap belt before the rover went out of control, and had hit her head on the console in front of her when they impacted the bottom of the gorge. A nasty bruise had formed on her forehead.

Dr. Wong left the cab to get his kit, and performed a quick examination of Lafrenière upon his return. "No sign of spinal injury…" He pulled up both her eyelids in turn, and shone a light into them.

"I think she has a concussion," Dr. Wong concluded. "Minor, perhaps."

"Then it's not so bad?" Christopher asked.

The physician shook his head. "No, not really. But I'd like to give her a checkout on the compudoc when we get back to base camp."

"Let's get her to a bunk," Christopher said. "Ollie, we're still in line-of-sight. Get on the horn to Darmstadt. Let them know what happened."

As Wong and Christopher carried the unconscious Lafrenière out of the cab, Solovyov went to the right-seat comm station and sat down. He tapped a set of commands on the touchscreen, but instead of a prompt to begin transmission, the system crashed. He tried re-initializing the console, and the same thing happened. Finally, he ordered a self-diagnostic, but even that did not come up.

By now Christopher had returned to the cab, leaving Dr. Wong to tend to Lafrenière. He walked up to the comm station and stood behind Solovyov, his hands grasping the seat back. "Problem?"

Solovyov nodded. "The high-gain antenna appears to be off-line."

"What's wrong?"

"I am not certain. Not even the BITE is responding," Solovyov said, referring to the antenna's built-in test equipment.

"Try the low-gain. Relay through base camp."

Solovyov nodded, keying in another sequence. Once again, the system would not respond.

"What's going on?" Christopher asked. He thought for a moment, then answered his own question. "There was a loud bang. Is it possible the external comm pallet was damaged?"

"It is possible," Solovyov nodded.

"All right. We'll go outside and check it out. Worse comes to worst, we'll contact Darmstadt when we get back to Lunae." Gesturing at the impact pattern on the windshield, he asked, "What about that?"

"I do not believe that is a problem," Solovyov said. The duraplex was impregnated with a dye that became visible when critical stresses incurred, but the canopy—designed to withstand moderate impacts—was still transparent. "However, we may lower the cabin pressure to –"

He was abruptly interrupted by a warbling tone as the master alarm sounded again. They looked at each other, then bounded to the engineering console. Christopher got there first. The commander deactivated the klaxon, which automatically brought up the relevant status window on the screen.

"Ollie! We're losing oxygen!"

"Switch to the redundant unit!" Solovyov exclaimed.

Christopher's hands flew over the touchscreen. A schematic flow diagram depicting the plumber's nightmare of lines and pipes in the rover's underbelly appeared on the monitor. "Backup tank on-line...What the hell? We're still losing oh-two!"

"What's with the master alarm?" Wong demanded as he tumbled into the cab. "What's going on here?"

Solovyov shouldered Christopher out of the seat and took his place. "It is not in the tanks," he said quickly. "It is in the main distribution manifold."

"Can you bypass it?" Christopher asked as he glanced at the oxygen status line. It was reading 31%, and falling rapidly.

"I am attempting to do so now," Solovyov said tersely as he entered new commands. The oxygen reading was now 19%.

"Oh, hurry," Wong pleaded as the numbers began dropping to single digits.

"I think we've got ourselves a problem," Christopher sighed.

Solovyov was still typing when the count reached zero.

Christopher patted the engineer on the shoulder. "It's OK, Ollie. You did your best."

"What'll we do now?" Wong asked in a tone of virtual despair.

Solovyov stared at the console. He wanted to slam his fist in frustration. Then it suddenly occurred to him that the rover had one other source of oxygen. The engineer entered a new set of commands, and as he did so, the low rumbling of the motor diminished to silence.

"I have taken the engine off-line," Solovyov explained. "I am now tying in the oxidizer tanks to the liquid flash evaporator, and redirecting the output to environmental control."

"How long will that give us?" Christopher asked.

"The system was not designed for this, so there will be a continuous 64.9% loss to the outside. However..." Solovyov finished typing, "I would estimate...approximately five hours."

Christopher spoke aloud Solovyov's next thought. "But it's going take us at least 18 hours to get back to Lunae."

"And just *how* are we supposed to get back to base camp anyway?" Wong shouted. "We don't have five hours if we bring the engine back on-line!"

"Electrolyzing our water supply will give us another half an hour, perhaps more," Solovyov suggested.

Wong looked on the verge of tears. "That's...that's not enough!"

"So, we have a little over five and a half hours of air," Christopher said. "It's a start. Let's build on it." He looked in turn at Solovyov and Wong. "Come on, I want ideas. How can we stretch that?"

The engineer was silent, but Wong opened his mouth as though he wanted to say something. Then he stopped, as though he'd thought better of it. He changed his mind again, and his mouth proceeded to open and close several times in succession like some fish in distress.

"Doc," Christopher allowed himself a grin. "If you have something to say, I want to hear it."

Wong nodded. "I think…" he paused, and swallowed deeply. "I think— I believe…" He summoned up some spine and finally managed to blurt it out. "If, there is—*isn't* enough, for all four of us…we will, we'll have to leave someone behind."

Christopher's fleeting smile disappeared instantly. "That is not an option!" he exclaimed.

Wong cowered, and his head seemed to shrink into his chest like a tortoise. Christopher took a deep breath, and his sudden flash of fury vanished as quickly as it had erupted.

"It's not an option," the commander repeated, a little more calmly. "Geez, Doc, you're talking nonsense. *Think* about it! Whether there's three or four bodies isn't going to make one hell of a difference. Besides…there are some people itching for an excuse to kill Ares. One of us dying would fit the bill nicely." He looked at Wong and Solovyov again. "Come on, guys—*think! I want some ideas!*"

It is said that when a person is facing certain death, their entire life flashes before their eyes. But Solovyov recalled only one incident. His first job out of university had been with NPO Precision Instruments. The young graduate's pride was matched only by his desire to impress himself upon the veteran engineers of the bureau. But on the morning of his first day the alarm did not go off, and he was two hours late for work. The manager of his division glared at him with his arms crossed sternly across his chest…much as Christopher was doing now. Embarrassed and ashamed, Solovyov had muttered a string of apologies for his tardiness, vehemently promising never to be late for anything again.

"Late," Solovyov murmured aloud.

"What?" Christopher asked.

"Late." Solovyov looked Christopher in the eye. "The Ares 5 ERV. For the next mission. It was launched almost two months late."

Dr. Wong and Christopher stared at him silently. Not waiting for a reply, Solovyov got up and went over to the comm station. He activated the console and displayed the most recent Ares itinerary uplinked from Mission Control.

"The Ares 5 Earth Return Vehicle was scheduled to be launched at the same time as our departure in preparation for the next mission, but it was delayed by a mechanical difficulty."

"Where is it going to land?" Christopher asked, excitement creeping into his voice.

Solovyov scrolled down the data until he found the information. "Latitude 22.3 degrees north, longitude 85.1 degrees east. Near the eastern rim of Fesenkov Crater."

"When?"

"It is scheduled to aerobrake into parking orbit in 5 hours, 17 minutes."

"Still too far for us to drive, but..." Christopher shook a finger at the screen. "If...if the ERV was brought down here instead, our problems would be over!"

"Yes."

"All right, Ollie!" Christopher patted the engineer on the shoulder. "Get suited up, and see what you can do about getting our communications back."

The three men went to the rover's airlock. Christopher and Dr. Wong ran through a checklist to ensure that nothing was missed and helped Solovyov into his spacesuit. Once he was fully geared, the engineer and the commander switched roles. Solovyov and Wong assisted Christopher into most of his own suit—except for the gloves and helmet. Christopher would monitor the progress of the EVA from inside the rover, but he would be ready to assist at a moment's notice. Both he and Dr. Wong kept their helmets beside them so they could talk to Solovyov through the built-in short-range radios.

Solovyov locked down his helmet and pressurized his suit. After a leak check and a radio test, Christopher handed him a standard repair kit. Then he entered the airlock and cycled it. Unlike the spacesuits of the early space program, the ones used by the Ares astronauts were high-pressure models that eliminated the need to pre-breathe oxygen. The throbbing sound of the pump diminished to silence as the chamber approached the ambient pressure of the Martian atmosphere outside. Finally a green indicator illuminated the wall panel. Solovyov opened the outer hatch and stepped outside.

The Sun was red and small, already low in the dust-dimmed sky as nightfall approached. Long shadows were being cast by the boulders, and as he surveyed the landscape a terrible irony struck him. There was plenty of oxygen locked up as mineral oxide in the endless store of ruddy rocks that surrounded the rover, but they had neither the equipment nor the energy to extract it.

Immediately to the right of the airlock hatch was a narrow ladder that allowed access to the roof of the rover. Solovyov held his kit in his right hand while his left steadied him on the ladder. After just a few careful steps up the ladder, he could peer over the top of the vehicle. Immediately in front of him was the rover's scientific instrument bay. Currently empty, it had been designed to support experiments mounted outside that needed direct exposure to the Martian environment. Just beyond the bay, at the rear of the vehicle, was what was left of the external comm pallet.

He keyed his radio. "Solovyov to Commander Christopher."

"Yeah, Ollie," the commander acknowledged from inside the rover. "How's it looking?"

Solovyov clinically outlined what he saw. There may have been a mathematical term to describe the shape the high-gain radio dish was in, but he didn't know what it was. It no longer even remotely resembled a parabola. Dark scorch marks emanated from the point of impact, an intricately mangled mess of broken metal that had once been the rover's external comm pallet. The low-gain antenna was nowhere to be seen.

"Is it beyond repair?"

"Yes. It looks like it took a direct hit from a meteorite. It is completely destroyed." Solovyov had the spare parts to repair individual components, but nothing to replace the entire system. "I am coming down," he announced as he descended from the ladder and returned to ground level.

Dr. Wong's voice appeared on the loop. "What'll we do now?"

"Think of something else," Christopher replied. "Any ideas, Ollie?"

Solovyov stared out toward the horizon. Mars was about half the size of Earth, so the horizon looked closer than what he was used to. Could a solution be just as near? Both the rover's high- and low-gain dishes were totally inoperative. If these were beyond repair, there was no way they could contact Earth—*unless…*

A sudden inspiration hit Solovyov, but he dismissed it immediately. There was one very, very slim possibility, but it was far too outrageous for him to mention aloud.

Christopher saved him the trouble. "Say, what about…the Viking 1 Lander?"

"What?" Dr. Wong stammered.

"Ollie, would it be possible to use the high-gain dish on the Viking?" Christopher asked hopefully.

"Possible," Solovyov said hesitantly. "But —"

"Good!" Christopher cut him off. "What would it take to get the unit up?"

Before Solovyov could reply, Dr. Wong cut in. "Why, that's...that's outrageous! That machine is decades old! It'll never work. I'd get a better response doing CPR on a corpse."

"We got data back from Voyager until a couple of years ago," Christopher pointed out. Designed to explore the outer Solar System, the twin spacecraft had been launched in 1977. "So, Ollie, how about it? What do we need?"

Solovyov furrowed his brow in thought. "Judging from my earlier examination of the spacecraft, it appears to be in excellent condition. I did not remove any components from the antenna mechanism or the communications subsystem, so only its age would be of concern. Very well. I have a standard kit with me now. We will also need...lubricants, hydraulic fluid...a power pack...more spare components—see if you can find an extra kit, and a tablet to interface with the Viking's electronics. Oh, of course—take out a receiver from one of the spare spacesuits."

"All right, I'm going to get that stuff now. I'll be coming out in a few minutes. Christopher out." Solovyov heard a soft thump as he put his helmet down before the channel clicked off.

"Oleg, this idea is sheer lunacy," Dr. Wong whispered. "Tell me honestly. What are its chances?"

"Not very good," Solovyov said bluntly. He believed it was an understatement.

While he waited for Christopher to join him, Solovyov decided to perform a quick walk-around inspection of the rest of the rover. He made his way to the front end of the vehicle, careful to avoid the edge of the gorge. The area was crumpled up like aluminium foil, and the impact on the windshield looked even more menacing than it did from the inside.

"Just went to download some stuff from the computer," Christopher came back on. "I've put on my gloves and helmet. Be out there real soon."

Solovyov didn't know what he wanted from the computer, and didn't bother to ask. As he made his way around to the starboard side of the vehicle, he spotted something silvery among the rusty red rocks. He took a few steps toward the anomaly and identified it. It was the rover's low-gain antenna, blown off the comm pallet and deposited on the ground, the small dish now crushed like a pressed metallic flower.

By the time he returned to the airlock hatch, he could see Christopher's spacesuited figure emerging from it. Once he was out, the commander reached back into the compartment to get the equipment.

"Hey, Ollie. Give me some help here."

Christopher handed Solovyov a tablet, a second standard repair kit, some spools of wire, and other assorted spare parts. His own load consisted of a few small canisters of hydraulic fluid and graphite-based lubricant as well as the spare receiver. The two men walked away from the rover until they were out far enough to see the roof. Then they stopped and looked up. Solovyov pointed, and Christopher whistled softly.

"You know, Ollie, I don't think we're going to be able to fix that."

"I had already come to that conclusion."

"That's, uh…that's pretty bad," Christopher observed. "Looks like E.T. is going to have to find a new way to phone home."

"What?"

"Forget it. Let's get going."

Before setting off for the Viking site, both men topped off the oxygen supply in their spacesuits at the EVA replenish station near the rover's airlock.

"Listen, Doc," Christopher said. "Without the rover's antennas that little helmet of yours isn't going to pick us up beyond about…" He turned to Solovyov.

"Eight hundred metres," the engineer answered.

"You got that, Doc? Eight hundred metres. I want you to try and monitor our progress from the roof camera. If we need you out here and we're out of range, I'll raise my arms like this." Christopher walked in front of the cab window and demonstrated the gesture.

Solovyov saw Dr. Wong nod behind the duraplex and reply "Understood" verbally. He glanced at the watch built into the left wrist of his space-suit. "The ERV will aerobrake into parking orbit in 4 hours, 51 minutes. Commander, we must hurry. We are losing precious time."

"Ollie…"

But Solovyov had already broken into an awkward run toward the Viking 1 Lander, which was now just over a kilometre east of the rover.

"Ollie! Wait, slow down! Don't need to run!"

Solovyov ignored him. It was difficult, with all the equipment he was carrying, but he managed a well-paced jog. The light Martian gravity helped tremendously. Time was of the essence. The chances of re-activating the old Viking were slim at best, but it was their only hope for contacting Mission Control before the rover's oxygen was depleted. Even a few minutes could turn out to be crucial. Why, Solovyov wondered, could Christopher not understand this? Sometimes, the man's infernal cheeriness was more than enough to –

Something seemed to grab his right foot, and the rusty red ground rushed up to meet him.

"Ollie!" Christopher exclaimed as Solovyov fell, his shout loud enough to distort his voice over the suit radio.

The light gravity saved Solovyov once again. It gave him just enough time to let go of his tools and put his hands in front of him to break his fall. His gloves hit the ground, his elbows buckled—and his faceplate stopped within a centimetre of a large, jagged red rock.

Solovyov moaned and rolled onto his side.

"Ollie!" By now, Christopher had reached the engineer. "Are you OK?"

"I...I believe so."

"Can you stand?"

Solovyov tried, and stopped instantly as pain radiated from his foot. He gingerly attempted to move it again, but the pain only increased as he did so. "My foot," he gasped. "My right foot. I think...I have sprained it."

"Doc!" Christopher barked. "I think we need your services already."

"OK...I'm almost, already partially suited, so –"

"No!" Solovyov tried to wave him off, a ridiculous gesture the doctor could not have possibly seen. "He must stay with Dr. Lafrenière. Help me up, Commander. There is no time!"

For a moment, Christopher stood in motionless indecision. Finally he said, "All right, stay put, Doc." He went about picking up Solovyov's tools, clipping what he could onto his utility belt and carrying the rest under his left arm. Next he retrieved the tablet, verified that it was not damaged by the fall, and put it inside his right thigh pocket.

The commander took Solovyov's left arm and brought it over his shoulder. He wrapped his right arm around the engineer's waist. Slowly, the two men came to a standing position, then they began the long walk to the Viking 1 Lander. Hobbling along like two drunks in their bulky spacesuits, they made painfully slow progress. For Solovyov, the emphasis was on the painful part.

The radio came on. "Wong to Christopher."

"Go ahead, Doc."

"Michèle is coming around. I think...to be all right...keep an eye on..."

"I'm glad to hear it, Doc." Christopher tapped the side of his helmet as the signal faded in and out before finally vanishing. "Doc?"

"We are moving out of range," Solovyov stated. They had no time to stop and listen, and Solovyov wasn't sure he wanted to hear any more of the doctor's depressing banter anyway.

"Well, I hope he doesn't drive off with the rover."

"Commander!"

"Just kidding."

When they finally arrived at the site of the old spacecraft, Solovyov glanced at his watch noted that the ERV would enter orbit in four hours and thirty-six minutes. A walk that should have taken nine or ten minutes had taken almost half an hour. Solovyov silently cursed himself for his clumsiness.

Christopher noted with obvious relief that the Viking had escaped damage in the meteorite shower. Solovyov cursed again, this time at whatever deities that might exist for choosing now of all times to punish his atheism by leaving the old Viking unmolested for decades while the Ares 4 rover was hit twice in one day.

Christopher and Solovyov made their way to the Viking's north-western face, the side opposite the soil sampler arm. They found an electronics compartment which would give them access to the communications subsystem, opened it, and got to work. Despite Christopher's help, their progress was hindered by Solovyov's injury. Neither of them carried anything that could be used as a walking stick, so the engineer had to keep one hand on the lander to steady himself.

As the astronauts worked, night fell over Chryse Planitia. Like on Earth, the Sun set in the west, behind the two men. Unlike on Earth, the thin atmosphere diffused little light, making Martian sunsets short, but spectacular in their own way. As the Sun dipped below the horizon, the salmon-pink sky gave way successively to a brief orange dusk, a dim crimson radiance, and finally a soft rosy afterimage. Blackness rapidly flowed into the vacancy left by the retreating glow.

But the astronauts witnessed none of this. Their only response was to turn on their helmet beacons. Solovyov turned briefly to look back at the rover, and noticed that none of its external lights were on.

"The ERV should have aerobraked into orbit by now," Christopher announced. "Assuming it did, it'll have to begin its deorbit burn in…fifty-two minutes to make a landing here."

The exposed electronics compartment was now connected by a spaghetti-like mass of wire to the tablet and receiver clipped to the side of the spacecraft. Solovyov blinked to clear a bead of sweat that had found its way into his eye. He had known from the outset that reactivating the long dormant lander, assuming it was possible, would be a time consuming process.

But the Viking was in excellent condition, and their task was relatively straightforward. All they needed was time. Solovyov glanced at his watch, and willed himself not to slam his fist in frustration. They were not going to get it.

"Hey!" Christopher said sharply. "What's the problem, Ollie?"

Solovyov gritted his teeth in frustration. "In order to land here, the ERV must execute its deorbit burn in 24 minutes."

There was nothing more to say. At the current position of Mars in its orbit about the Sun, it would take at least 25 minutes for a radio signal to be received on Earth and a reply sent back.

"We are out of time."

After a short silence, Christopher said, "But the ERV can receive S-band transmissions through its contingency low-gain antenna. So, what's there stopping *us* from trying to land the beast here *ourselves*?"

"It is possible," Solovyov conceded. "But we need the command sequence to –"

"It's on the tablet, in the folder SCOUT."

Solovyov raised his eyebrows. "How did you know…Scout?"

"Sure. Weren't you ever a Boy Scout? You know, 'Be Prepared'? I figured –"

Solovyov waved him into silence as he continued working with renewed determination. "You impress me, Commander."

"Well, we have 19 minutes until the ERV is in range." Christopher became instantly serious again. "In order for it to land here, we must uplink the sequence as soon as it is, because it'll have to begin its deorbit burn two minutes later."

"The timing is very delicate."

"I know. But we must press on. We'll finish."

Somehow, they did. Hydraulic lines were filled, leaks were found and plugged. A portable power pack was installed to provide electricity, as the output of the Viking's own radioisotope thermoelectric generators had decayed below its threshold level years ago. Since they were now to transmit data instead of voice, the spare spacesuit receiver was disconnected. Finally, Christopher boosted Solovyov up to lubricate the lander's high-gain antenna dish.

"The ERV is almost in range," Christopher announced. "Are we ready?"

Solovyov did a quick survey of their work. It was an undignified mess, but all seemed to be in order. "I believe so."

"All right then. Move the dish into position."

Solovyov entered the command into the tablet's keypad. He waited a moment for the program to take effect. He waited a little longer.

Absolutely nothing happened.

"Ollie…"

"I do not understand!"

"We're *really* running out of time this time, Ollie. I mean it."

Solovyov frantically checked the makeshift connections: Wiring, voltages correct…hydraulic pressure acceptable…

"Two minutes. Are you sure we did everything right?"

"Yes!" he exclaimed adamantly.

The commander exhaled deeply. "Then there's only one thing left to try." With those words, he took a step back from the lander, and Christopher—in his youth a running back for Creighton University—raised his right foot and delivered a prodigious kick to one of the Viking's silvery terminal descent propellant tanks.

"Try it now."

Solovyov took a second to recover from his amazement, then punched the command into the tablet again.

The Viking 1 Lander's high-gain antenna came to life.

"Works every time."

"Must have dislodged some dust…" Solovyov muttered by way of explanation.

"Whatever." Christopher looked at his watch. "That's it. We are coming in range, standby…now! Punch it, Ollie!"

The engineer re-initialized the tablet and entered the command to activate the high gain antenna once more. The Viking's parabolic dish responded, turning toward the eastern horizon and lowering itself to an angle of twenty-one degrees above horizontal.

"I have acquired the ERV's transponder signal. I will now uplink the new command sequence." Solovyov loaded the program into memory, then entered the code word KVUGNG to initiate its transmission to the spacecraft approaching overhead.

"It is done," he said at last.

"Is there any way to know if…"

"No. There was no time to echo the sequence back." No way to know whether the software patch was installed correctly. Solovyov looked at his watch. "If it was successful, the ERV will initiate its deorbit burn in twenty-eight seconds, and will begin its descent two minutes, four seconds after that."

There was nothing more to do but wait. An uneasy silence fell between the two men. Solovyov wanted to return to the rover to await the ERV's landing, if it ever came. His foot was killing him. But Christopher made no move to leave.

Solovyov cleared his throat, and decided there was no better time. There was something he wanted to get off his chest.

"Commander Christopher."

"Yes?"

"There is something I must tell you."

"Shoot."

"I do not like to be called 'Ollie'."

A pause. "Oh," Christopher said at last. Then, "Well, to tell the truth, I never liked the way you call me 'Commander'. It's too rigid, too formal. I don't like rigid and formal. So, how about we both quit while we're behind and call it even?"

"That would be acceptable," Solovyov said simply.

Over a minute had elapsed since the command sequence had been transmitted to the ERV. When Christopher spoke again, Solovyov expected an order to return to the rover. Instead it was, "Hey, turn off your light."

"What?"

"Turn off the light," he said and did just that.

Puzzled, Solovyov followed suit. His eyes adjusted quickly to the darkness, and he looked up into the night. The spectacle filled him with wonder.

A global environmental restoration effort coordinated by the space-based Earth Observing System had helped clear terrestrial skies, but no starscape on the home planet could match what the two men were seeing now. They were enveloped in an eternal ebony blanket, studded by crisp, jewel-like points of light which shone with their full intensity. It was almost as if nothing stood between the astronauts and the cosmos. Almost, but not quite, for Mars still had a tenuous atmosphere. Although the stars barely flickered, each bore a faint pinkish tint bestowed by the dust perpetually aloft in the Martian sky.

"You know," Christopher said, "when people start living here permanently, I think someone will have to change the words to 'Twinkle Twinkle Little Star'."

"I know about that!" Solovyov exclaimed.

To the south was Phobos, the larger and nearer of Mars' two natural satellites. It shone as a pale oval in the heavens, about two-thirds the apparent diameter of the familiar Moon.

"Hey, look at that!" Christopher was facing the west, and his finger pointed upward.

Solovyov followed his arm into the sky. He squinted and a few seconds later, he could see it too. A star, a little brighter than the rest, was moving—rapidly—down toward the horizon.

"Deimos?" Christopher asked sceptically.

"No. It is too bright, and moving too quickly." For the first time, Solovyov felt hope swell within him. "I believe it may be the ERV."

"You're right!"

As the two astronauts watched, the point of light briefly became a very short, intense streak before finally disappearing below the horizon.

Christopher whooped and jumped, hanging in the air for a moment before coming down. "Right on, right on!" He punched Solovyov's shoulder. "We did it, Oleg!"

"The ERV *will* touch down in thirty-one minutes," Solovyov said with a hint of a smile—an expression Christopher couldn't see.

The engineer turned to gaze upon the faithful old Viking. He could just barely make out the tiny Mutch Station plaque. Its words could not be read in the darkness, but its meaning had never been more clear. This monument was more than a tribute to one person. It was a symbol of the spirit, drive, and dedication of humanity to spread forth from its African cradle, cross the continents, sail the oceans, take to the sky, and soar beyond.

When next humans walked these rusty red plains, perhaps they would make them their home. On that day, Solovyov hoped, a more fitting memorial would be erected at this place, and people would come to Chryse Planitia to marvel at an artefact of the past—and of the future.

Oleg Solovyov faced the Viking 1 Lander again, and in a gesture of unmitigated sentimentalism, saluted.

Afterword

"Dedication" is very special to me because it was my first professionally published science fiction story. As I recalled in the Preface, I wrote "Dedication" while a student in the engineering science programme at the University of Toronto, and I entered it into a then-new contest called the Isaac Asimov Award for Undergraduate Excellence in Science Fiction and Fantasy Writing. Now called the Dell Magazines Award, the contest was established by *Asimov's Science Fiction* magazine and the International Association of the Fantastic in the Arts to encourage emerging student writers of speculative fiction. All these years later, it is still difficult to describe my shock at becoming the first winner of the Dell Award in the first year it was offered. "Dedication" was subsequently published in the November 1994 issue of *Asimov's*. I will forever be grateful to Rick Wilber, Sheila Williams, and the late Gardner Dozois for launching my writing career.

The story was inspired by a painting called "Visit to Utopia" by Pat Rawlings, which depicts a pair of astronauts approaching the Viking 2 Lander at Utopia Planitia on Mars. One of the astronauts is holding a flag but its nationality is obscured because the scene is backlit by the Sun—a thought-provoking ambiguity. I got an autographed print of "Visit to Utopia" at a Mars conference of the American Institute of Aeronautics and Astronautics, and the framed print still hangs on my wall. This was the first professional scientific conference I had ever attended, and it was made possible by a student essay contest sponsored by The Planetary Society. A congratulatory letter signed by Carl Sagan remains one of my most treasured possessions.

My original intention had been for the fictional Ares 4 crew to visit the Viking 2 Lander as depicted in "Visit to Utopia", however, during the course of my research I learned about planetary scientist Tim Mutch who was the lead investigator for the Viking surface imaging team. Mutch disappeared in 1980 during a descent from the peak of Mount Nun in the Himalayas, and to honour his memory NASA renamed the Viking 1 Lander at Chryse Planitia (the Plains of Gold) the Thomas A. Mutch Memorial Station. A plaque with the inscription as described in the story remains on display with a Viking Lander engineering model at the Smithsonian Air and Space Museum, waiting to be brought to Mars and attached to the real thing. This was the origin of the title "Dedication" and the opening line, "There is no gold on the Plains of Gold."

Viking was one of the most ambitious NASA space missions of the 1970s. A pair of spacecraft, each consisting of an orbiter and a lander, were launched in 1975 and arrived at Mars the following year. Perhaps the highest profile science investigations were the astrobiological experiments on the landers that were intended to detect microbial life on Mars. Some have argued the ambiguous results of these experiments killed interest in Mars for years. Indeed, more than a decade would pass before the launch of the next Mars missions, the ill-fated Russian Phobos 1 and 2. The space agencies of the world have become more careful with respect to setting public expectations about life on Mars.

The Ares mission scenario described in the story is the Mars Direct architecture devised by Robert Zubrin and his colleagues in the 1990s.[1] "Dedication" depicted the original version of Mars Direct that called for dual launches of an Earth Return Vehicle (ERV) and a Mars Excursion Vehicle (MEV) at every 26-month launch window to Mars. To address concerns about launch vehicle capability and the need to store large quantities of hydrogen on the Martian surface, Zubrin and his colleagues developed a modified architecture called Mars Semi-Direct that splits the payloads over three launches instead of two. In Mars Semi-Direct, the first launch delivers a

---

[1] Zubrin, Robert and Wagner, Richard. *The Case for Mars*, Touchstone, 1996.

Mars Ascent Vehicle (MAV) to the surface where it makes methane propellant with stored hydrogen reacted with carbon dioxide from the Martian atmosphere. A second launch delivers an ERV that does not land but remains in orbit around Mars. Finally, the astronaut crew takes the third launch aboard an MEV-like habitat that lands near the now fuelled MAV. When their surface mission is complete, the crew takes the MAV into Mars orbit and rendezvous with the ERV for the flight back to Earth. Mars Semi-Direct became the basis of a number of subsequent studies including a series of NASA Design Reference Missions.[2]

To my knowledge, "Dedication" is one of the earliest depictions of Mars Direct in science fiction. The mission architecture has appeared in a number of novels including *The Martian Race* by Gregory Benford (Warner Aspect 1999), *Return to Mars* by Ben Bova (Harper Voyager 2000), and *First Landing* by Robert Zubrin (Ace 2001). Mars Direct was also mentioned in a 2004 episode of the American TV drama *The West Wing*. "Dedication" also has what I believe to be one of the first depictions of using an old Mars probe as an emergency communications system. The protagonist in the 2000 movie *Red Planet* uses the Mars Pathfinder for this purpose, as does the Mark Watney character in the novel and film *The Martian* (Random House 2014). *Star Trek* fans may note that the Shaun Christopher character is an homage to the original series episode "Tomorrow Is Yesterday" which established that Christopher survives his Ares 4 adventure and goes on to command the first successful human mission to Saturn.

Shortly after I wrote this afterword, a trio of international robotic missions arrived at Mars: the Al Amal orbiter of the United Arab Emirates, the Tianwen-1 orbiter and lander of China, and the Perseverance rover of the United States. For the moment, a human mission to Mars remains the realm of science fiction. In the meantime, both The Planetary Society and the Dell Magazine Award continue to inspire the next generation of explorers and dreamers. For more information, please visit www.planetary.org and www.dellaward.com.

---

[2] Salotti, Jean-Marc and Heidmann, Richard. "Revisiting Mars Semi-Direct", 9th IAA Symposium on the Future of Space Exploration, 7–9 July 2015.

Raise the Nautilus

South Pacific Ocean
1,700 nautical miles east-northeast of New Zealand
June 1916

The being that emerged from the depths of the sea was humanoid, but it did not look human. The creature's tough beige hide was the texture of elephant skin. The head was a heavy bronze sphere, its face a cyclopean eye criss-crossed by a metallic grate. A pair of thick tubes protruded below the glass eye, snaking back to a heavy cylinder on its back.

Commander Thomas Jennings watched as the diver was hoisted onto the deck of the research ship RRS *Discovery*. Two sailors helped the diver to a seated position on a wooden bench, where they began to disconnect the tubes and unfasten the helmet. A third man in a brown trilby and civilian attire, tall and gaunt with a sharp nose, stared impassively at the scene.

Jennings turned and looked out to sea. A few hundred yards distant the grey bulk of his ship, the cruiser HMS *Euryalus*, drifted serenely on the sparkling waters. The tranquillity was a welcome relief from the fierce storms that had delayed the start of the operation for days.

The sailors lifted the helmet off the diver, revealing the ruddy face of a middle-aged man with sweat-soaked dark hair. He closed his eyes, threw his head back, and inhaled deeply.

"Are you all right?" Jennings asked.

The diver, Jonathan Badders, nodded. "Yes, sir. Thank you."

© The Author, under license to Springer Nature Switzerland AG 2020
E. Choi, *Just Like Being There*, Science and Fiction,
https://doi.org/10.1007/978-3-030-91605-3_2

Donald McCabe, a civilian from the Meta Section of the Directorate of Military Intelligence, stepped forward. "What did you see? Is she there?"

Badders nodded again, this time smiling. "She's there. I saw her. The *Nautilus*."

* * *

As the executive officer of HMS *Euryalus*, Commander Jennings was glad to be back aboard his ship. Both *Euryalus* and *Discovery* had been launched in the same year, but the latter seemed much older. *Discovery* was a wooden three-masted auxiliary steamship, the last of her kind to be built in the British Empire. The Admiralty had purchased *Discovery* from the Hudson's Bay Company, refitting the cargo ship to serve as a floating base for Operation Mobilis—the Royal Navy's attempt to raise the *Nautilus*.

Jennings, Badders, McCabe, and the divisional officers gathered around a table in the commanding officer's day cabin.

"What's the state of the *Nautilus*?" asked Captain Richard Powell.

"She's at a depth of forty-one fathoms, with a slight list to starboard of about seven degrees," reported Badders. "Her stern is buried in about eighteen feet of hard clay, but otherwise she appears to be undamaged."

"When can the salvage operation begin?" asked McCabe.

"Weather permitting, as early as tomorrow," said Lieutenant-Commander Eugene Seagram, an officer from the Royal Navy Engineers who had been seconded from the Admiralty to support Operation Mobilis.

"About bloody time," said McCabe. "The Smith-Harding report was quite specific about the last known location of the *Nautilus*. I can't believe it took almost a month to find her."

Jennings, Powell, and the key divisional officers had been briefed by McCabe on the *Nautilus* file just before their departure from Auckland. Of particular interest to the War Office was a description in the Smith-Harding report, corroborated by earlier accounts from Aronnax, of "a destructive weapon, lightning-like in its effects" that could stun or kill men. His Majesty's Government was still telling the public that the Great War was going well, but military men like Jennings knew the terrible truth. McCabe's impatience was annoying but understandable. Such a weapon, in the hands of the British, could break the stalemate on the Western Front.

"Badders' report on the condition of the *Nautilus* is excellent news," said Seagram. "It means we can proceed with the original salvage plan with little modification." He spread across the table a schematic diagram of the submarine, copied from a trove of documents seized five years ago during a joint raid by the Directorate of Military Intelligence and the British Army on the ancestral palace of the late Prince Dakkar in the Bundelkhand region of India. "The five pontoons from the *Discovery* will be deployed as follows: Three above the stern, and two above the bow. For additional buoyancy, we will run two hoses down to blow the main ballast tank of the *Nautilus*. The biggest challenge will be the unforeseen need to tunnel the clay under the stern to place the harness and lifting chains for the aft pontoons."

"How long will this take?" Powell asked.

"Three weeks," Seagram replied.

Donald McCabe rolled his eyes and shook his head.

"Very well," Powell said. "Get some rest, gentlemen. We have a big day tomorrow."

There was one other aspect of the Smith-Harding report that had made an impression on Commander Jennings—the fanatical hatred of Prince Dakkar, later known as Captain Nemo, for the British Empire. How ironic it would be if Nemo's invention ended up saving it.

* * *

Donald McCabe's cynicism was vindicated. It actually took fifty days for Badders and his team of divers to pass the harnesses and lifting chains under the *Nautilus*, attach the pontoons, and connect the hoses to the main ballast tank. Just tunnelling the clay under the stern took the entirety of the originally-estimated three weeks.

But at long last, everything was ready. Jennings, McCabe, and Seagram returned to the *Discovery* to supervise the operation.

"Proceed with blowing the stern pontoons," Seagram ordered.

Jennings watched as the pumps roared to life, the needles on gauges began to move, and the hoses snaking into the water stiffened.

"We're starting with the stern first," Seagram explained, "which will hopefully avoid any centre-of-gravity issues that might arise if both ends were lifted at the same time."

Euryalus was anchored a few hundred yards away. She was joined by HMRT *Rollicker*, an Admiralty tug that had been dispatched from Auckland six days ago.

"There!" Yeoman Farley called out excitedly, pointing to a coloured float that had bobbed to the surface.

"I see it," Seagram said. "That's good. It means the stern has been lifted off the sea floor. Stop the air to the aft pontoons, and start blowing the forward pontoons!"

The men scanned the surface of the water with binoculars, looking for the second coloured float. It did not appear.

Seagram glanced at the pressure gauges and frowned. "The bow should be lifting by now." He shook his head. "It's taking too long. Start the tertiary pump. We'll try blowing the main ballast tank on the *Nautilus*."

After ten minutes, the second float finally appeared. Moments later, the surface of the ocean began to froth violently, like water in a saucepan brought to boil.

"Something's wrong," said Jennings.

Suddenly, both of the bow pontoons popped to the surface. The massive cylinders, twelve feet in diameter and thirty feet long, briefly cleared the water before crashing back down in a torrent of spray.

Seconds later, it was the *Nautilus*.

Jennings gasped as the bow of the submarine came up, smashing into one of the pontoons and tossing it aside. She rose like a giant breaching metallic narwhal, the armoured steel spur that protruded from her nose glistening in the midday sun. Then, with a great splash, the *Nautilus* slipped back under the waves and quickly disappeared from view.

* * *

"What the devil went wrong?" Captain Powell demanded.

"It appears the centre-of-gravity of the *Nautilus* is further aft than we expected," Seagram explained. "When we blew her main ballast tank, the momentum caused the bow to rise so fast it slipped out of its harness, separating it from the pontoons and sending everything to the surface. The impact with the pontoon dislodged the hose and opened the ballast tank vents, allowing water back into the main ballast tank and submerging the *Nautilus* again."

"How could you have not known the location of the centre-of-gravity?" McCabe asked. "We have the drawings of the *Nautilus*."

"We have the *design* drawings, not the as-built drawings," Seagram explained. "The mass properties of the final build appear to differ significantly from the original design."

"What do we do now?" Powell asked. "The damaged pontoon is beyond repair."

Jennings was in a glum mood. The Admiralty was demanding daily wireless reports and bristling at the delays and costs, even threatening to cancel Operation Mobilis. He thought for a moment, and then said, "I have an idea."

"Commander?"

Jennings spread a nautical chart onto the table. "The only reason this operation was even feasible is because the depth of the ocean here is unusually

shallow. There was an island here until it was destroyed in the volcanic eruption of 1882. We know from the Fessenden oscillator soundings we took
during the search phase that there are even shallower areas nearby." He traced
a finger on the chart. "For example, just two nautical miles due east of our
present location, the depth is only sixteen fathoms."

"Yes!" Seagram exclaimed. "I see where the XO is going with this." He
produced a notepad and began to sketch. "We can use the four remaining
pontoons, not to bring the *Nautilus* to the surface, but to lift her just high
enough off the sea floor for the *Rollicker* to tow her to the shallower area,
and then we deliberately ground her there. Her ballast tank vents will have
to be repaired, but it will be much easier and safer for Badders and his divers
to work at the shallower depth. We'll then use the shorter hoses from the
Discovery that we couldn't use before to blow both the main and secondary
ballast tanks, and we can also use the trim tanks to compensate for the centre-
of-gravity issue. Taken all together, there should be sufficient buoyancy to
bring the *Nautilus* back to the surface."

"You think this is feasible?" Jennings asked.

"I'll need to work out the details with my team," Seagram replied, "but
yes."

"How long until we're ready for a second attempt?" Powell asked.

"Three weeks," Seagram said.

McCabe shook his head. "Is that your answer to everything?"

* * *

The preparations took almost triple the time. Just lifting the *Nautilus* off
the seafloor took three weeks, followed by two days for the *Rollicker* to tow
and ground her in the shallower water, and finally another five weeks for the
divers to repair the ballast tank vents, reattach the pontoons, and connect the
additional hoses and lines.

On the day of the second salvage attempt, it appeared as if every man
aboard *Euryalus* had gathered along the portside railing to watch. For a
moment, Jennings had the amusing and absurd thought that the ship might
very well tip over.

Once again, the surface of the ocean started to foam and bubble, and
then the great bulk of the *Nautilus*, this time with pairs of massive pontoons
attached at deck level to bow and stern, burst through the frothing water.
Pressure-relief valves on the pontoons and the submarine opened, sending
up spray that shrouded the *Nautilus* in rainbowed clouds of vapour. Slowly,
the *Nautilus* righted herself and settled onto the surface, her bobbing motion
sending small waves towards *Euryalus* and *Discovery*.

For a moment, nobody could speak. The men of the *Euryalus* just stood there, mesmerized by the sight. And then, a massive cheer erupted through the crowd.

"About bloody time," muttered Donald McCabe.

The real *Nautilus* bore little resemblance to the version depicted in the popular press during Captain Nemo's infamous reign of terror in the late 1860s, with its ridiculous serrated fins, bulbous viewports, and outsized rivets. The real *Nautilus* was two hundred and fifty feet in length from the tip of the armoured spur at her bow to the large propeller and fish-like rudder at her stern. A pilothouse protruded about a third of the way along her deck, below which were mounted large hydroplanes for diving control. She was a sleek grey war machine, in some ways a long-lost ancestor of the Royal Navy's new K-class submarines that were only now commencing sea trials.

Jennings turned to Seagram and shook his hand. "Congratulations!"

"Sir!"

Jennings made his way through the crowd to find Jonathan Badders and the divers, shaking each of their hands in turn. "Well done, gentlemen. Yours was the most vital and dangerous job. This accomplishment belongs to you."

"Thank you, Commander," said Badders.

It took a day to run the towing hawser between the *Nautilus* and the tugboat *Rollicker*, and the following day the flotilla commenced its long journey back to Auckland under the escort of *Euryalus*. The most difficult part of Operation Mobilis was over, but there was still one more task to perform.

* * *

Only one body had been found aboard the *Nautilus*, and what to do with it became a rather delicate question.

"Sir, you can't be serious!" exclaimed Jennings, sitting across the table from Powell in the captain's day cabin.

"I am quite serious, and my decision is made," said Powell. "Captain Nemo is to be reburied at sea with honours."

"Sir," Jennings pleaded, "that man was an enemy combatant. All those ships he sank. I would not be surprised if there are men right here on the *Euryalus* who knew some of the people he murdered."

"I'm aware of that," Powell said. "Nevertheless, Captain Nemo was a comrade of the seas, and arguably he was even a subject of the British Empire for a time. Like it or not, he is entitled to the honour."

"I don't like it, sir."

"Your objection is noted," Powell said. "However, if it makes you feel better, think of it this way. What better revenge could we exact than to bury him under the flag of his adversary?"

Jennings thought for a moment and decided that he had greatly misjudged his commanding officer's sense of humour.

So it came to be that Jennings, Powell, and the divisional officers of the *Euryalus* found themselves in dress uniform standing at attention before the Union Jack draped coffin of Captain Nemo. Jonathan Badders and five of his fellow divers served as pallbearers, lifting the coffin and slowly marching towards the edge of the deck.

"The sea is the largest cemetery, and its slumberers sleep without a monument," said Captain Powell. "All other graveyards show symbols of distinction between great and small, rich and poor. But in the ocean cemetery, the king, the clown, the prince, the peasant, the hero, and the villain are all alike, undistinguishable."

The pallbearers set the coffin down on a platform at the edge of the deck, and they removed and folded the flag. Badders pulled a lever, the platform tilted downward, and Captain Nemo returned to the sea.

Commander Jennings kept his face neutral and respectful, but in his mind's eye he was smiling. Captain Nemo had been taken from his beloved *Nautilus*, which in turn had been taken from him. His final resting place would be a remote and obscure part of the South Pacific, amongst the sharks, alone and far from the coral cemetery of his crewmates. The Royal Navy had given Captain Nemo exactly what he deserved.

It was a very British thing to do.

* * *

At full speed, *Euryalus* could have made it back to New Zealand in about four days. But after a week at sea, the flotilla was still more than three hundred nautical miles out from Auckland. The problem was the antiquated *Discovery*, which could barely manage eight knots under sail and steam. Jennings was still making daily wireless reports to his impatient masters in the Admiralty, and he knew that every additional day at sea was another day that something could go wrong.

On the eighth day, his fears were realized.

"What is it, XO?"

"We have company," Jennings said, handing his binoculars to Powell.

The captain looked out the bridge windows of the *Euryalus* to where Jennings was pointing. "Are we expecting any Allied shipping in this area?" Powell asked.

"No, sir," said Jennings.

"I'm not aware of anything either," said McCabe.

Jennings spoke into a voice pipe. "This is the bridge. Can we get a range on the target bearing zero-four-nine?" After a few moments, he bent his ear to the pipe. "Probably about thirteen nautical miles," he reported, "but it's hard to tell with the mist. Wait a minute." He held up a finger. "Lookout reports the target appears to be turning. The rangefinder will need a moment to re-establish coincidence."

"I don't like this," Powell muttered.

"The mist is clearing, and the target is no longer turning," Jennings reported. "Sir, the rangefinder has a new solution. Eleven nautical miles, and closing rapidly."

Powell turned to McCabe. "If that's a German warship, how soon until we're in range of her guns?"

"For their Pacific fleet, I would expect it to be a light cruiser like ourselves, with similar weaponry," McCabe said. "Main guns would have a range of around 15,000 yards."

After a few minutes, Jennings raised his binoculars again. "Target now at ten nautical miles, still closing fast. I can see a profile and battle ensign." He flipped through the pages of the silhouette book. "Confirmed, she's German sir! Most likely SMS *Scharnhorst*, an armoured cruiser."

The klaxon sounded, and Powell spoke into a voice pipe. "All hands, this is the Captain. Action stations, action stations. Inbound hostile vessel. This is not a drill." He turned and called out to a young man. "Yeoman Farley!"

"Yes, sir!" said Farley, pad and pencil ready.

"Get on the wireless. Make to *Rollicker* and *Discovery*. 'German cruiser inbound. *Euryalus* engaging. *Rollicker* and *Discovery* to maintain current speed and heading.'"

"Yes, sir!"

"Make course zero-one-five to bring the target within the firing arc of the fore-turret," Powell ordered. "Prepare to fire when the *Scharnhorst* breaks seven nautical miles."

From the bridge windows, Jennings could see the turning of the fore-turret and the single barrel of the 9.2-inch Mark X gun beginning to rise. The approaching German cruiser was now visible without binoculars.

"Seven nautical miles!" Jennings called out.

At that moment, a flash and a puff of smoke erupted from the *Scharnhorst*. Seconds later, the unmistakable screech of a shell in flight could be heard, followed by a concussive boom and an explosive geyser of foamy white seawater a few hundred yards off the bow of *Euryalus*.

"Return fire!" Powell ordered.

The *Euryalus* trembled as her forward gun discharged. A plume of water erupted short of the *Scharnhorst*.

"Visual splash spotters, recalculate the solution," Jennings ordered. "Gunnery, open fire when ready."

Euryalus fired again, and this time, the shell found its target. On the *Scharnhorst*, an explosion blossomed amidships.

"Hit, sir!" a lieutenant exclaimed.

Jennings bent his ear to a voice pipe, then turned and peered through his binoculars. "The *Scharnhorst* is no longer closing. She appears to have stopped. However, her guns seem to be –"

A shell exploded a few yards off the bow.

"– very much intact, sir!"

Yeoman Farley returned with a piece of paper. "Message from the *Rollicker*, Captain!"

Powell read the telegram, his eyes wide. "*Rollicker* reports spotting another ship bearing three-four-nine, identified as SMS *Gneisenau*."

Jennings called into a voice pipe. "I need confirmation and range to the new target, bearing three-four-nine." Moments later, he said, "Second target confirmed, Captain! Range five-and-a-half nautical miles."

Suddenly, a massive explosion rocked the *Euryalus*, throwing the men to the deck. Powell got to his feet and staggered to the engine order telegraph. "Helm, make course three-five-zero, half speed. I want to engage with a broadside."

The helmsman rotated the steering stand, but the *Euryalus* did not turn.

"I have a report from damage control," Jennings said. "We've been hit astern, probably by a torpedo. Rudder and propeller are damaged. Helm will not respond."

"Bring the aft-turret to bear on the *Gneisenau*," Powell ordered. "Fore-turret, keep firing on the *Scharnhorst*." He looked around. "Where's Yeoman Farley?"

"Here, sir!"

"Make to *Rollicker* and *Discovery*. 'If Germans attempt to intercept or board, make no attempt at armed resistance.'"

"Sir?" said Farley, incredulous.

"Young man, we are in a terrible tactical situation," Jennings explained. "*Discovery* and *Rollicker* cannot run, and we cannot protect them against two cruisers. Their only chance is to stay out of the fight—and pray that these particular Germans are inclined to abide by the London Declaration." He jerked his thumb. "Now, go!"

"Wait!" Powell said suddenly. "XO, what's the depth of the ocean here?"

Jennings consulted a nautical chart. "About 880 fathoms."

"Yeoman!" Powell thought for a moment, then said, "Further message to the *Rollicker*. 'Cut pontoons. Scuttle the *Nautilus*.'"

Farley's eyes widened. "Sending this right away, sir!"

McCabe approached. "Captain, I assume you want me off the bridge?"

Powell nodded. "Get below, Mr. McCabe."

The fore- and aft-turrets fired simultaneously. Another explosion blossomed on the *Scharnhorst*, this time at the bow.

Moments later, a blast ripped through the *Euryalus*.

Jennings struggled to his feet. He touched his forehead and felt blood. "Captain? Lieutenant? Is everyone all right?" Mumbled affirmations drifted across the shattered bridge. Staggering to the smashed forward windows, he looked out and beheld a horrific sight. The fore-turret had been hit, reduced to bent and twisted metal from which smoke poured out of serrated holes.

Hearing a shuffle of feet, Jennings turned to see Captain Powell at his side. They looked at each other for a moment, and both knew what needed to be done. They could not run, and it was only a matter of time before the fire ignited the magazine of cordite charges. When that happened, the *Euryalus* would explode in a fireball.

"All hands, this is the Captain. Abandon ship. I repeat, abandon ship. All hands to the lifeboats immediately." Powell turned to Jennings. "Divide the decks between us. Check as many compartments as you can."

"See you at the lifeboat, sir!"

Jennings raced through the smashed and smoke-filled decks and compartments of the *Euryalus*, calling out and directing dazed crewmen, both able-bodied and injured, to external hatches and lifeboats. The men were afraid but not panicked, the evacuation carried out with a stoic British efficiency that Jennings would later recall with tremendous pride.

The last place Jennings checked was the wireless cabin. He had expected it to be already evacuated, but was surprised instead to find an unconscious Yeoman Farley sprawled on the deck. Relieved to find Farley alive and breathing, Jennings hooked his elbows under the armpits of the unconscious man and lifted him in a fireman's carry. His lungs burned from smoke and exertion, but finally he managed to stagger outside.

Captain Powell, Lieutenant-Commander Seagram, and the divisional officers of the *Euryalus* were already in the last lifeboat. With their assistance, Jennings brought Farley aboard and then hopped inside himself. The derrick swung the lifeboat over the side and down to the water.

"Is everyone accounted for?" Jennings asked. The pained, vacant look in Powell's eyes was the only response he got.

They began to row furiously, joining the other lifeboats in trying to get as far away from the *Euryalus* as possible. Suddenly, there was an explosion. The occupants of the lifeboat turned to watch an ominous mushroom cloud rise into the air. But it wasn't the *Euryalus*. It was the *Scharnhorst*.

Jennings knew *Euryalus* would soon share the same fate, but for now he took grim satisfaction in seeing the German ship go down first. For the moment, *Euryalus* was still afloat, as were the apparently undamaged *Discovery* and *Rollicker*.

And the *Nautilus*.

"Sir!" Jennings tapped Powell on the shoulder, and pointed.

The *Nautilus* was still on the surface, still attached to its four pontoons, still hooked up to the *Rollicker*. A short distance away was the second German cruiser, the *Gneisenau*. Jennings watched in stunned disbelief as a boatload of seebataillon, marines of the Imperial German Navy, came alongside the *Rollicker* and boarded. About twenty minutes later, smoke began to billow from the *Rollicker*'s funnel and the tug began to move, with the *Nautilus* in tow and the *Gneisenau* following behind.

Under different circumstances, the men in the lifeboat might have shouted and cursed. But nobody spoke. Cold and in shock, and physically and mentally exhausted by the battle, the harrowing escape, and the imminent loss of their ship, they just sat in helpless silence as the Germans sailed away with their prize.

London, England
December 1916

A note on the menu reminded customers to limit their order to two items for lunch and three for dinner. Commander Thomas Jennings flipped through the pages but decided he wasn't hungry. He put down the menu and sat back, waiting.

Half an hour later, a tall man wearing a brown trilby exited the War Office Building across from the restaurant. Jennings stood, remembering to push his chair under the table, and walked across the street to intercept.

"Mr. Jennings," said Donald McCabe. "What a surprise to see you."

"You're a hard man to reach," Jennings said. "I've been trying to contact you for well over a month."

"What do you want?"

Jennings gestured. "Let's talk a stroll through the gardens."

The two men walked down Horse Guards Avenue, then turned to pass through the black wrought iron gates into Whitehall Gardens.

"The ocean is big, but the German Pacific fleet is not," Jennings said. "Yet somehow, we managed to run into two German cruisers less than a day out from Auckland."

"Perhaps the ocean isn't as big as you think."

Jennings stopped and turned. "And why was the order to scuttle the *Nautilus* not carried out? Or was the order never sent? I found Yeoman Farley unconscious in the wireless room."

"As I recall, we were being shelled and torpedoed by German cruisers. Men tend to fall down under such circumstances."

"This is all a game to you DMI people, isn't it?" Jennings hissed, his anger rising. "Let me remind you of the cost. The *Euryalus* destroyed. Ninety-three injured, many seriously. Twenty-eight missing and forty-four dead, including Jonathan Badders. The crew of the *Rollicker* taken prisoner. And the *Nautilus* in the hands of the Germans."

"Forgive the pun, Jennings," McCabe said with condescension, "but as a Navy man, you're out of your depth."

"Then explain it to me."

McCabe lit a cigarette. "Would it interest you to know that German war production is down almost nine percent from this time last year?"

Jennings was puzzled. "What does this have to do with anything?"

"Kaiser Wilhelm fancies himself a man of science," McCabe continued, "but in reality he is quite mad. He is obsessed with the *Nautilus* and her weapons. Well, let him have it. An armoured battering ram is hardly impressive in the age of torpedoes. As for this lightning weapon, I suppose what the Germans would call a blitzwaffe…"

McCabe flicked ash onto the grass. "He can have that, too. Directed energy weapons are years if not decades away from widespread practical military deployment. This war will be won with more ships, more artillery, more guns, perhaps more aeroplanes – not with fantastical inventions from the realm of scientifiction. In the meantime, if it pleases the Kaiser to pursue this folly, every pfennig spent on this fantasy is a pfennig that is not being spent on real weapons that will make an actual difference to the outcome of this war."

Jennings was silent for a moment, contemplating McCabe's words. At last, he said, "I think you're wrong."

"I don't care what you think."

"Then maybe you should care that I know something about you," Jennings said.

"And what would that be?" McCabe asked.

"That your mother was born in India."

McCabe stared at Jennings for a moment, then threw the cigarette butt to the ground and crushed it with his shoe. "It goes without saying that everything I've told you is sensitive and privileged information. I should remind you that the penalties for disclosure under the Official Secrets Act are rather severe."

"So are the penalties for treason," Jennings hissed.

A thin smile crossed McCabe's lips. "We each have to fight the war in our own way." He tipped his hat. "Goodbye, Mr. Jennings. God save the King."

The man from the Directorate of Military Intelligence walked away. Commander Thomas Jennings made no attempt to follow. Alone amongst the greenery of Whitehall Gardens, he wondered when the history of the Great War is written whether Operation Mobilis would be remembered as a success or failure. And by whom.

Afterword

"Raise the Nautilus" was the closing story in an anthology called *20,000 Leagues Remembered* that was released on 20 June 2020 to commemorate the 150th anniversary of the classic Jules Verne novel *20,000 Leagues Under the Sea* (*Vingt mille lieues sous les mers*). Within the pages of the novel the *Nautilus* submarine was described by Verne in exquisite detail, and in many ways the vessel could be considered a character in its own right.

There is an erroneous notion perpetuated by the overrated 1954 Disney film adaptation that the *Nautilus* was nuclear powered. In fact, the *Nautilus* of the novel was powered by chemical batteries, specifically, a hugely scaled-up version of a zinc-carbon primary battery called a Bunsen cell, which was invented in 1841 by the German chemist Robert Bunsen (perhaps better known for the "Bunsen burner"). The Bunsen cell used a carbon cathode in nitric acid and a zinc anode in dilute sulphuric acid with a porous separator between them. Verne knew that the Bunsen cells of his day would not be adequate to power the *Nautilus*, so he hypothesized an improved version in which the zinc is replaced with sodium extracted from seawater. It is fortunate that Captain Nemo's *Nautilus* operated in the world of science fiction, because the reaction of metallic sodium with sulphuric acid would in fact have generated a great deal of energy in a rather explosive manner.

Certainly the *Nautilus* would have needed every joule of energy it could get, as its performance described in the novel met or even exceeded the capabilities of modern submarines. In the novel, the highest speed achieved by the

Nautilus was 40 miles per hour (35 knots) while running under Antarctic ice, and at one point Captain Nemo boasted to Professor Aronnax that the vessel could reach a top speed of 50 miles per hour (43 knots). The fastest military submarines ever built were the Cold War era Soviet Lira (Лира) class nuclear-powered attack submarines that had a submerged top speed in excess of 40 knots. Equally remarkable was the capacity of the batteries aboard the *Nautilus*. At one point in the novel, Professor Aronnax estimated that the *Nautilus* had been sailing continuously for fifteen or twenty days. One of the most advanced modern diesel-electric attack submarines is the German Type 212 class (U-Boot-Klasse 212), which is equipped with an air independent propulsion (AIP) system using proton exchange membrane hydrogen fuel cells that allow the vessels to stay submerged for up to three weeks. However, the Type 212 has a submerged top speed of only 20 knots.

As described by Captain Nemo in the novel, the *Nautilus* was a double-hulled vessel with T-shaped irons connecting the inner and outer hulls. The outer hull consisted of steel plates two inches (five centimetres) thick, which is similar to the thickness of the steel pressure hulls of many modern submarines. There was no indication of the thickness of the inner hull. The maximum depth reached by the *Nautilus* was 16,000 yards (14.6 km), which was achieved when Captain Nemo ordered the vessel to the bottom of the ocean in the South Pacific. Pressures at this depth would have far exceeded the structural integrity of the *Nautilus*, and indeed the depth exceeds what is now known to be the lowest point of the seabed, the Challenger Deep of the Mariana Trench in the western Pacific Ocean, which has a depth of 10.9 km. It is a remarkable fact that humans have spent more time on the surface of the Moon than the bottom of the Challenger Deep.

It would not be until after the Second World War that submarine design gradually approached the performance of Verne's *Nautilus*. Admiral Hyman Rickover of the United States Navy was an early proponent of nuclear propulsion for naval vessels, and in 1954 the U.S. Navy commissioned the USS *Nautilus*, the world's first operational nuclear-powered submarine. In February 1957, the USS *Nautilus* logged her first 60,000 nautical miles, bringing to reality the 20,000 league distance achieved by her fictional name-sake. USS *Nautilus* successfully completed the first submerged transit of the North Pole in August 1958, and later took part in the American naval quarantine of Cuba during the missile crisis of October 1962. The submarine was decommissioned in 1980 and put on display at a naval museum in Gorton, Connecticut with recognition as a U.S. National Historic Landmark.

Operation Mobilis was based on the real-life salvage of the USS *Squalus*, an American diesel-electric submarine that sank during a test dive off the coast of

New Hampshire in May 1939. Twenty-six sailors were killed, but the lives of the remaining thirty-two crewmembers and one civilian were saved over the course of a thirteen hour rescue operation using a diving bell called a McCann Rescue Chamber. The U.S. Navy then undertook a long and difficult salvage operation over the course of the next four months in which the *Squalus* was eventually raised and towed to the Portsmouth Naval Yard. Following extensive repairs, the submarine was recommissioned as the USS *Sailfish* and went on to serve in the Pacific theatre during the Second World War.

The title and theme of the story were influenced by the 1976 Clive Cussler novel *Raise the Titanic*, in which a team attempts to salvage the ocean liner and recover a substance that could tip the balance of power during the Cold War. The revelation at the end of the story that Donald McCabe had deliberately allowed the *Nautilus* to fall into the hands of the Germans with the intention of diverting resources from their war effort is based on the history of the Nazi V-weapons programme in the Second World War. The V-1 flying bomb and V-2 ballistic missile were terrible weapons that killed thousands of people (including 20,000 slave labourers who were forced to build them), but in the end these weapons not only did not help the Nazis win the war but arguably contributed to their defeat. Hitler's obsession with the futuristic V-weapons diverted enormous resources away from other military needs that would have had a much greater impact on the war, such as more long-range bombers to continue attacking Britain or more fighters to defend German cities from Allied bombing. According to the historian Daniel Todman, "Proportionate to the two countries' sizes, the rocket programme cost Germany at least as much as the Americans spent on the project to build the atomic bomb, but with very much less result...In retrospect, the German rocket programme was, by a distance, the greatest waste of resources by any combatant country in a supremely wasteful war."[1]

---

[1] Todman, Daniel. *Britain's War: A New World, 1942–1947*, Oxford University Press, 2020, pp. 359–360.

The Son of Heaven

Tsien Hsue-shen was born in 1911, in the last weeks of Chinese imperial history, and at 23 travelled to the U.S. to study aeronautical engineering at the Massachusetts Institute of Technology. Preferring theory to the practice that MIT then emphasized, he soon moved to Caltech and began to follow a path that would lead to his becoming one of the most eminent rocket scientists in the U.S.
—Aviation Week & Space Technology, 2007 Person of the Year

Pasadena, California
September 1950

They came for him in the late afternoon.

The two agents from the Immigration and Naturalization Service strode up the walk to the small one-storey redwood clapboard and brick house at the end of East Buena Loma Court. Arriving at the front door, one of the agents knocked.

Jiang Ying opened the door, her baby daughter Yung-jen in her arms.

"Is this the Tsien residence?"

"Yes."

The agents flashed identification. One produced a piece of paper.

"We have a warrant for the arrest of Mr. H.S. Tsien."

Jiang Ying silently stepped aside. The baby began to cry.

The INS agents entered the house and surveyed the small, sparsely furnished living room. In the corner, at the foot of a bookshelf, two-year-old Yucon cowered, his eyes wide.

"Mr. Tsien!"

Tsien Hsue-shen appeared, his expression resigned. His hands were clasped over his stomach, as if nursing a wound.

"Please come with us."

Shanghai, China
August 1935

As the steamship *President Jackson* pulled away from dock, Tsien watched the crowd at the pier recede into the distance. Tsien Chia-chih, his father, and Chang Langdran, his mother, were still waving. The name they had given him, Hsue-shen, meant "study to be wise", reflecting all the hopes they had for him.

The *Jackson* turned for open water, picking up speed. Tsien took a deep breath and thought about the irony of his good fortune: his post-graduate education in America was made possible by a conflict between China and the United States.

Under the terms of the 1901 peace treaty following the Boxer Rebellion, the victorious foreign powers imposed reparations against China. But the American share of the indemnity turned out to be twice the amount of actual U.S. claims. President Theodore Roosevelt decided to return the surplus by establishing the Boxer Rebellion Indemnity Scholarship, allowing the best and brightest Chinese students to study in the United States.

Tsien looked out to sea, to the new ocean before him.

California Institute of Technology
March 1937

"...and the older woman says, 'There's a terrible curse attached to this diamond.' So the young woman asks, 'What curse?'" Paul Epstein paused dramatically. "The older woman lowers her voice and says, 'Mister Plotnick.'"

Theodore von Kármán shared a laugh with his colleague from the Physics Department. The two professors were in von Kármán's office, taking a break from grading undergraduate exams.

"And when will *you* be cursing some unfortunate woman, Theodore?" Epstein asked.

The old aerodynamicist grinned mischievously. "I never found the need to."

There was a knock at the door.

Von Kármán looked up. "Oh, come in, Hsue-shen."

Tsien walked into the office, dressed in a suit and tie as he always did on campus. He was a short man, with a smooth round face that looked younger than his twenty-six years. He parted his thin black hair awkwardly on one side.

"Professor Epstein," Tsien said, standing straight. But when addressing his doctoral supervisor, he bowed slightly. "Von Kármán lǎoshī." He handed over a stack of papers. "Here is the numerical analysis of the transfer functions you asked for."

"Thank you, Hsue-shen. I have something for you as well." Von Kármán rummaged through the journals and papers scattered across his desk. "You mentioned that you want to learn about rockets." He pulled out a journal, folded it open to a page, and handed it to his student.

Tsien read the names of the authors. "Frank Malina and William Duncan Rennie."

"Bill is in Canada, visiting his parents," said von Kármán. "But Frank is around. You should meet him."

"Thank you, von Kármán lǎoshī." Tsien turned and left the office.

Epstein looked at von Kármán. "What did he call you?"

"It means 'old teacher'," von Kármán explained.

Epstein understood. "He respects you greatly."

"The feeling is mutual."

"Tsien was in my relativistic quantum course," Epstein continued. "He is brilliant."

"Yes, he is good," von Kármán said. "The other Chinese students have a nickname for him: 'The Son of Heaven'."

"Tell me," Epstein said with a twinkle in his eye, "Do you think he has Jewish blood?"

Arroyo Seco, California
October 1937

They were called the Suicide Squad.

Lead by Frank Malina, with the sponsorship of von Kármán, they were a mad monk outfit of Caltech graduate students and local enthusiasts that conducted rocket experiments. After an unfortunate incident in the Gates Chemistry Building, the group was exiled from campus and forced to continue their work at Arroyo Seco, a dry river bed canyon a few miles from Caltech.

Malina, Tsien, and Rennie, along with a chemist named John Parsons and a mechanic named Edward Forman, had worked until 3:00am to prepare a rocket motor for their latest test. After catching a few hours of sleep, they drove at dawn to Arroyo Seco and mounted the motor onto its test stand. They connected the fuel and oxidizer lines, then piled sandbags around the apparatus before retreating to their viewing site.

Malina handed Tsien the trigger. "Will you do the honours?"

Tsien pressed the button. The rocket motor ignited, bright flame leaping from the nickel-steel nozzle into the early morning air.

The five young men cheered.

Rennie checked his stopwatch. "Forty-four seconds and counting. That's a record, guys!"

Parsons pointed. "What's that other flame there?"

"I believe the fuel line has broken," Tsien said calmly. "It is on fire."

Malina's eyes widened. "Uh, guys…*Run!*"

The Suicide Squad fled across the canyon, moments before the rocket engine exploded behind them.

Pasadena, California
December 1938

Tsien was in a foul mood.

Last night, he had gone to see *The Adventures of Robin Hood*, taking a break from the stress of finishing his PhD thesis. A patron at the theatre had demanded the usher eject Tsien from his seat. He did not want to sit next to an Asian.

Someone knocked gently on the corner of his desk. Tsien looked up. It was Frank Malina, his tall, lean frame towering over the desk. His angular face, topped by a short crop of curly dark hair, sported a razor-fine moustache.

"Are you all right?" Malina asked.

"Yes."

Malina looked sceptical, but continued. "Hey, do you know Sid Weinbaum?"

"No."

"Sid's one of the research assistants in the Chemistry Department. He's throwing a party at his place tomorrow tonight. Do you want to go?"

The following evening, Tsien and Malina found themselves strolling up the walk to Sidney Weinbaum's small grey bungalow on Steuben Street. Inside the house, some twenty or thirty Caltech students were sprawled out on the furniture and chairs of the living room. Tsien and Malina had come neatly dressed in vests, ties, and polished shoes—a dignified contrast to the sloppily attired Bohemian crowd around them.

Normally a loner at social events, Tsien found it surprisingly easy to talk with this group. He found sympathetic ears for his outrage at the recent Japanese atrocities in Nanjing. They discussed other international crises, including the Great Depression and the rise of fascism in Europe. Someone suggested Tsien should read the works of John Strachey.

Later, after refreshments were served, Tsien found himself talking to an attractive young blonde, telling her about his idea for a transcontinental rocketliner that could travel from New York to Los Angeles in an hour. Even trips to the Moon, Tsien told her, might be possible in the near future.

The young woman listened for a while, then smiled politely and excused herself. Perhaps she thought the strange little Chinaman had been drinking too much.

Caltech
November 1943

Blackboards went around all three sides of the lecture hall. Tsien had already filled two of them with his small, precise handwriting, and was well into the third, the chalk making gentle squeaks as he wrote.

"Professor Tsien, I don't understand the third equation on the second board."

Tsien continued to write without responding.

Moments later, another voice called out. "Sir, are you going to answer his question?"

"That was a statement of fact, not a question."

Finally, with the third blackboard filled, Tsien completed the last equation with a flourish and put down the chalk.

A timid hand was raised. "Sir, this method you have here, using the calculus of variations – is it fool-proof?"

Tsien gave the student a cold stare. "Only fools need fool-proof methods."

"Sir, we've had no quizzes, no midterms, no problem sets. Can you at least tell us something about the exam?"

"If you understand everything, you will be fine." His patience exhausted, Tsien turned and strode briskly out of the lecture hall.

A few days later, Tsien was called into von Kármán's office.

"Hsue-shen, a number of students have come to me expressing…concerns about your class." Von Kármán had lost weight and appeared to be in poor health. "You might consider changing your approach."

"Von Kármán lǎoshī, we are not teaching kindergarten. This is graduate school!"

"From my experience," von Kármán continued, "a good lecture is when one-third of them understand what you are talking about, a third has a pretty good idea, and the rest have no clue."

Tsien shook his head. "I am interested only in lecturing to people who understand everything."

"I know you prefer to do research," von Kármán said, "but as a professor you must recognize that teaching is also an important part of your responsibilities."

Tsien nodded. "I will do better, von Kármán lǎoshī."

Mojave Desert, California
10 December 1944

The Private A rocket sat poised at the base of its inclined launch tower, an angular grey metallic truss that stuck out starkly against the beige desert floor. On the horizon, the sharp peaks of the Granite Mountains could be seen.

Tsien focused his binoculars on the Private A. The rocket was small, barely eight feet in length, with four stubby tail fins for stabilization. Its main engine was augmented with four solid propellant boosters, which together would deliver over twenty thousand pounds of thrust in less than one-fifth of a second at the moment of lift-off.

As Tsien lowered his binoculars, he marvelled at how far they had come since the crazy days of the Suicide Squad. Their experiments had eventually attracted the attention of the U.S. Army Ordnance Corps, which began funding Caltech to advance the development of long-range rockets. With von Kármán's endorsement, Tsien obtained a security clearance to work on military projects. A new institution, the Jet Propulsion Laboratory, was established to carry out the research.

"Any news on Theodore?" Frank Malina asked. Von Kármán had recently undergone surgery for intestinal cancer.

Tsien shook his head. "The operation went poorly. He is still not well."

The eyes of the Caltech engineers and Army personnel were focused on the distant rocket.

"Here we go," whispered Malina.

There was a flash of flame, a cloud of smoke and sand billowed out, and the Private A raced up the rails of the launch tower. It cleared the structure and streaked into the heavens, a small black cruciform soaring against a crystal blue sky.

The onlookers cheered and patting each other on the back. It was the first successful launch of a large solid-fuelled rocket in the United States.

Shanghai, China
July 1947

Jiang Ying sang like an angel.

Her powerful soprano voice soared from the stage to the highest rafters of the Lanxin Theatre. Dressed in an elegant silk qipao, her lustrous black hair gleamed like lacquer under the lights, accentuating her delicate cheekbones and unblemished skin. She was the most beautiful woman Tsien had ever seen.

This was his first visit back to China since he set sail on the *Jackson* twelve years ago. His mother had passed away in his absence, but his father was still alive, and Tsien had spent several weeks with him in Hangzhou.

He was now in Shanghai, and earlier in the day had delivered a keynote speech at his alma mater Jiao Tong University. It was at the dinner following his speech that Tsien was offered the presidency of Jiao Tong. The recital at the Lanxin was the final gift from his hosts.

The audience was on its feet before Jiang Ying's final note had faded.

Tsien bribed his way backstage and somehow managed to find his way to her dressing room. She actually came out to see him, thinking to indulge another autograph seeker. But Tsien had other ideas.

"Will you go out with me?" It was all he could think to say. He was, after all, just an engineer.

Jiang Ying was annoyed. It was by far the worst pick-up line she had ever heard. But she said yes.

Tsien Hsue-shen and Jiang Ying were married less than two months later.

Caltech
6 June 1950

It was raining the day the FBI came to Tsien's office.

"Can I help you?" Tsien asked.

"I'm agent Hanssen, FBI. This is agent Roberts." They flashed identification and sat without being invited.

"What do you want?"

"Let me get right to the point," Hanssen said. "Are you now, or have you ever been, a member of the Communist Party?"

In shock, Tsien was unable to answer for several moments. Finally, he said, "Absolutely not."

Roberts produced a picture. "Do you know this man?"

"Yes. That is Sidney Weinbaum."

"What is your relationship with Mr. Weinbaum?" Hanssen asked.

"He was a research assistant in the Chemistry Department. I used to go to his parties when I was a PhD student, but I have not seen him recently. How is he?"

Roberts leaned forward. "Mr. Tsien, are you aware that these so-called 'parties' were in fact meetings of Professional Unit 122 of the Pasadena Communist Party?"

"No!" Tsien exclaimed.

Hanssen produced a piece of paper. "This is a copy of a membership list dated February 1, 1938. Your name is on this list, associated with the alias 'John Decker'."

Tsien's face was ashen. "This is impossible!" he stammered. "I am not a Communist! I have never been a Communist! I have no idea how I got onto such a list. I have never heard the name John Decker."

The FBI men waited for Tsien to calm down. Then, Hanssen said, "Tell me about Weinbaum. Would you say he's a loyal American?"

Tsien struggled to answer. "I am an engineer, and as an engineer, the only yardstick I have to measure anything is data. Since data cannot be applied to such intangibles as a person's character or political beliefs, I cannot speculate on the loyalty of Mr. Weinbaum."

Hanssen and Roberts looked at each other, then closed their notebooks and stood.

"That will be all for now," Hanssen said. "Thank you for your cooperation."

Later that day, Tsien received a hand-delivered letter from the headquarters of the Sixth Army at the Presidio in San Francisco, informing him that the U.S. Government had revoked his security clearance.

12 June 1950

"What are you going to do?"

Von Kármán had never seen his former student so upset. Tsien was white as a sheet, his hands trembled, and his eyes appeared moist. He was struggling to maintain his composure and looked about to burst into tears.

In truth, von Kármán was not in much better shape himself. He had never fully recovered from his cancer surgery in 1944 and was forced to turn down a position to lead the Scientific Advisory Group at the Pentagon. Though his mind was ever sharp, his body had become thin and weak.

"I want to return to China" Tsien said.

"You must not do that!" von Kármán exclaimed.

"Why not?"

"Because you will immediately bring further suspicion upon yourself. It will make you look guilty."

"I am guilty of nothing," Tsien said, "but I cannot work without a security clearance."

"You cannot work on the military rocket programs without a security clearance," von Kármán pointed out. "Your theoretical studies and your teaching are not affected."

"In China —"

"What is there for you in China?" von Kármán interrupted. "Do you still have a chance at the presidency of Jiao Tong University?"

"The offer was withdrawn. The 'new administration' was suspicious of my wife's links to the Kuomintang." Jiang Ying was the daughter of one of Chiang Kai-shek's military advisors. Tsien snickered at the irony.

"When you were in China, you toured some of their universities," von Kármán continued. "What did you think of their research facilities?"

"It is obvious, von Kármán lǎoshī. There is nothing in China that matches what is in America."

"Then if you return to China, your days of ground-breaking scientific research are over," von Kármán warned. "China needs people to rebuild the country, not sit around thinking about space travel."

Von Kármán's tone softened. "Think about what you are doing, Hsue-shen. Do not make hasty decisions. This crazy business will pass. I have been in America over twenty years, and I have seen how quickly things can change."

Tsien looked at his mentor with gratitude. "I am thankful that you are still at Caltech to counsel me. I do not know what I would have done without your wisdom…perhaps something I might have regretted later." He bowed. "Thank you, von Kármán lǎoshī."

"You have called me that for twelve years, Hsue-shen, but I would be grateful if you could start calling me something else: Lǎo péngyou."

Tsien smiled. His old friend's pronunciation was excellent.

Los Angeles, California
Fall 1950

Following his arrest on 7 September, Tsien was incarcerated for two weeks at a federal facility on Terminal Island. Under the Subversive Control Act of 1950, Tsien was charged with failing to divulge his membership in the Communist Party when he re-entered the United States after his 1947 visit to China. He was eventually released on bail but was forbidden from traveling outside Los Angeles County.

Tsien's first deportation hearing took place on 15 November in a government building at 117 West Ninth Street in downtown Los Angeles. With his attorney Grant Cooper at his side, Tsien waited for the proceedings to begin. Presided by INS examining officer Albert Del Guercio, INS hearing officer Roy Waddell, and State Department observer Nick Di Carlo, the hearing began with an investigation of Tsien's background and recent activities.

"What were you doing in China in 1947?" Del Guercio asked.

"Visiting my father and getting married," Tsien deadpanned.

Later, two retired police officers named Hynes and Kimple took the stand.

"We have a membership list for the Los Angeles area Communist cells." Hynes handed a sheet to Del Guercio. "You can see that Mr. Tsien's name is on the list, next to the alias 'John Decker'."

"How was this list obtained?" Di Carlo asked.

"I'll ask the questions here," Del Guercio snarled. "Mr. Kimple, how did you get this list?"

"I was undercover," Kimple replied. "I had infiltrated the Communist Party as an assistant to their membership director. Every time a list or membership card came by my desk, I'd write down the names."

Grant Cooper leaned forward. "So this list is not an original. It's a copy in your handwriting."

"Yes," Kimple replied.

"And Mr. Tsien's name," Cooper continued. "Did you copy it from an actual membership card or from another list?"

Kimple paused. "I don't remember. Probably another list, maybe."

Cooper pressed. "Assuming another list, was it an actual membership list or a recruitment list?"

"What difference does that make?" Del Guercio asked.

"My client is a *rocket scientist!*" Cooper exclaimed. "It makes perfect sense for the Reds to try recruiting him, but that doesn't mean he ever joined them."

Del Guercio waved his hand impatiently. "Let's move on." He took a piece of paper from Waddell. "Mr. Tsien, you state in your deposition that you had no idea the gatherings at the Weinbaum residence were meetings of the Communist Party."

"That is correct," Tsien said.

"This is hard to believe, since you said yourself that you attended these events on a regular basis." Del Guercio put down the paper. "Is it *possible* that these *could* have been Communist meetings?"

"It is possible, but I had no way of knowing that," Tsien insisted. "They were always arguing about politics and world affairs. Sometimes they would ask for my opinions because I am Chinese. There were heated political discussions, but I thought this was part of the university experience in America."

"Mr. Tsien," Del Guercio asked, "in the event of a conflict between the United States and Communist China, would you fight for the United States?"

Cooper threw up his hands.

"What kind of a question is that?" Di Carlo interrupted.

Del Guercio glared. "Mr. Di Carlo, this is an INS hearing. You are here as a courtesy to the State Department, which insisted on your presence because of Mr. Tsien's unique background. So, kindly stick to observing and be silent about it!"

"Don't answer the question," Cooper said.

"Mr. Tsien, you will answer the question," Waddell ordered.

Tsien did not respond for a long time, until Del Guercio looked about to prompt him, and then he spoke. "My essential allegiance is to the Chinese people. If a war were to start between the United States and China, and if the war aim of the United States was for the greater good of the Chinese people—and I think it would be—then, of course, I would fight on the side of the United States."

Pasadena, California
30 April 1951

The dark Ford Tudor had been parked outside the Tsien household since the late morning. Tsien would periodically take a peek through the venetian blinds. The car's windows were tinted, making it difficult to see if anyone was inside.

"How come no light, bàba?" Yucon asked.

Tsien picked up the toddler and carried him to an armchair. He sat and reached for a scientific journal but didn't even have time to open it before the telephone rang. Fearful of waking the baby Yung-jen, Tsien immediately got up and answered the phone, with Yucon still in his arm.

"Hello?"

There was no response.

"Who is this?"

The caller hung up.

Tsien replaced the handset, summoning his will to keep from slamming it down in anger. He went to the window and peered outside. The dark sedan was gone.

Jiang Ying emerged from the nursery.

"Is Yung-jen still sleeping?" Tsien asked.

"Yes," she replied. "Who was that?"

Tsien handed Yucon to his wife. "I am going to see Professor von Kármán."

"Is he on campus today? You should call to check."

"There is something wrong with our telephone," Tsien said.

Tsien drove to Caltech and found von Kármán in his old office on the second floor of the Guggenheim Building.

"I do not know how much more I can stand," Tsien said. "I have not been able to do meaningful work for almost a year. I cannot answer questions from the JPL engineers about my own papers. I cannot go to conferences. I am not allowed to even go to the beach in Orange County!" Tsien gripped his hands. "And I am certain I am being watched."

"I don't doubt it," von Kármán said grimly. "Things have gotten crazier with the fighting in Korea. I read in the *LA Times* that the FBI has even been spying on the Chinese Hand Laundry Association."

Tsien signed. "I fear my days in America are numbered."

"We're not finished yet. Grant Cooper is still pursuing appeals, and Lee DuBridge has been campaigning tirelessly for us," said von Kármán, referring to the president of Caltech. "Lee had written numerous letters on your behalf. He is even planning a trip to Washington."

"I do not want to go back to China," Tsien said, "but I may have no choice."

"There are *always* choices." Von Kármán thought for a moment, then reached into a drawer and pulled out a letter. "This is from Frank Malina. He only recently took that job with UNESCO, but he's already thinking about leaving the United States and going to Paris to pursue his art."

"I am not an artist."

"No, of course not." Von Kármán tossed the letter on his desk. "Do you know who else was here last week? Bill Rennie. We met for dinner, and he told me about some interesting projects going on in Canada."

"What kind of projects?" Tsien asked.

"Oh, he mentioned something about a fighter plane and a new missile, and he said their National Research Council is upgrading their high-speed wind tunnels in anticipation of future projects." Von Kármán rubbed his chin. "Yes, it was all very interesting."

Shortly after returning home, a mailman arrived with a registered letter for Tsien. It was on INS letterhead, dated 26 April 1951. The letter stated that the INS had reached its decision and determined that Tsien was "an alien who was a member of the Communist Party of the United States and is therefore subject to deportation"

Los Angeles, California
November 1952

Grant Cooper's legal tactics and Lee DuBridge's eloquent letters, and even a small campus protest (rumoured to have been "encouraged" by von Kármán) were all for naught. In early November, Tsien's last appeal was denied.

"This is very bad," Cooper said grimly. "It means you can now be picked up at any time and placed into custody."

Tsien and von Kármán sat silently in Cooper's office, letting the attorney's words sink in.

"I'm sorry," Cooper said.

Tsien shrugged. "Our bags are already packed. We are ready to go to China at any time."

Von Kármán pounded his fist on Cooper's desk. "America or China. No, I do not accept it! This is *not* a binary condition." He thought for a moment, then reached for the telephone.

"No." Cooper pointed to the door. "Use the pay phone across the street. I'm sure this one's bugged."

Von Kármán returned a few minutes later. "I've invited Bill Rennie to join us for lunch."

Tsien, Cooper, and von Kármán met Rennie at a restaurant on East First Street. A large red vertical neon sign with the words "Far East Chop Suey" ran up the building's beige facade.

"This Chinese stuff is good," Cooper said, chewing a mouthful of chicken and bean sprouts.

Tsien grimaced. "It is not Chinese."

Von Kármán put down his fork. "Bill, you just got back from Canada?"

Rennie took a sip of water and nodded. "Yeah, spent some time with my parents, then went to a seminar at the University of Toronto Institute of Aerophysics."

"And what have our northern friends been up to?" von Kármán asked.

Rennie wiped his mouth. "Quite a few things, actually. They've started flight testing a subsonic fighter plane called the CF-100, and there are proposals floating around for a supersonic interceptor." He pulled out a notebook. "I met this really sharp kid at UTIA, one of the PhD students." He flipped to a page. "Gerald Bull, that's his name. He's doing aerodynamic modelling for a missile project called Velvet Glove."

Tsien's eyes widened. "A rocket?"

"An air-to-air missile for their fighters," Rennie said. "But there's talk of a sounding rocket program for space research."

Von Kármán turned to Tsien. "Are these projects of interest to you?"

Rennie held up his hands. "Before we get too far, I should point out most of this stuff is being done by industry, which might not be the best fit for Hsue-shen. But I have some contacts at the High-Speed Aerodynamics Laboratory in Ottawa. It's a great facility, and they need fluid dynamics people to support the Velvet Glove program."

"Canada..." said Tsien, his tone uncertain.

"It is a possibility, and I think you should give it serious consideration," von Kármán said. "The research facilities in Canada must be much better than in China, and I believe it would be advantageous for you to stay in a Western country. You don't want to close any doors."

Tsien recalled his master's studies at MIT and the harsh Boston winters. "Is it cold in Canada?"

"Oh, yeah," Rennie said.

Cooper shook his head. "The government will never go for this."

"But we haven't asked," von Kármán pointed out. "In any event, my lǎo péngyou must first make a choice."

Two weeks later, Tsien made his decision.

15 December 1952

Neither the INS nor the FBI returned Grant Cooper's calls. But one person did respond: Nick Di Carlo from the State Department, who flew out to Los Angeles to meet Tsien and Cooper in the attorney's office.

"Your proposal is interesting," Di Carlo said. "I'm going to bring it to my superiors in Washington."

Cooper looked surprised.

"The INS still wants to send me back to China," Tsien said.

"Yeah well, I like to think that we in State have a more... strategic view of things than some of the other three-letter Washington bureaucracies." Di Carlo leaned back in his chair. "I'm sure we could get the Department of Defence on side too." He pointed at Tsien. "I had a chat with Navy Secretary Dan Kimball about you a few days ago. Do you know what he said?"

Tsien shook his head.

"Kimball said he'd rather shoot you than let you go back to China."

"I assume he was joking," Cooper muttered without humour.

Di Carlo continued. "I can't promise anything, but I'll do my best to advocate for this."

"Thanks," Cooper said.

Tsien asked, "Why are you advocating our proposal?"

Di Carlo looked Tsien in the eye. "We Italians have a saying, Mr. Tsien. I'm sure you've heard it, the one about keeping your friends close?"

Tsien understood. "I believe Sun Tzu said it first."

Di Carlo rose from his seat. "Merry Christmas, gentlemen. Let's see what the future brings."

Malton, Ontario, Canada
4 October 1957

The afternoon of the Avro Arrow's rollout was cool and sunny. A steady stream of cars, buses, and limousines had been arriving at the A.V. Roe Canada aircraft company parking lot since noon, disgorging scores of people.

Tsien could feel the excitement as the crowd swelled into the thousands. On stage were a number of dignitaries, including Air Marshal Hugh Campbell, Minister of National Defence George Pearkes, some Royal Canadian Air Force vice-marshals, several Members of Parliament, and the senior management of the A.V. Roe Company, led by president Crawford Gordon.

Scanning the crowd, Tsien spotted his friend Jim Chamberlain, the aerodynamicist and chief of technical design for the Arrow. It was Chamberlain who had invited Tsien to the rollout. Their eyes met, and Chamberlain waved.

Tsien returned his attention to the stage as Defence Minister Pearkes took the podium.

"Fifty years ago, the great Canadian pioneer John McCurdy, who is with us on the platform today, flew the Silver Dart, the first aircraft in Canada, and in fact the first heavier-than-air plane to fly in the British Commonwealth. History recognizes that event as the beginning of Canada's air age.

"Today, we mark another milestone—the production of the first Canadian supersonic airplane. I am sure the historians of tomorrow will regard this event as being equally significant in the annals of Canadian aviation."

As soon as Pearkes finished his speech, he pulled a gold cord that symbolically opened immense blue-and-gold curtains stretched across the mouth of the hangar. On cue, a fly-past of CF-100 fighters swept overhead as a band struck up a fanfare, and with deliberate majesty the CF-105 Arrow was slowly towed out. The crowd broke into applause as the gleaming white delta-winged aircraft was revealed in the late afternoon sunshine.

This was the first time Tsien had seen an actual Arrow, having worked only on theoretical analyses and wind tunnel models at the High-Speed Aerodynamics Laboratory. Tsien's expert eye spotted the aerodynamic improvements to the nose and engine nozzles he had worked on with Chamberlain, drawing on his experience with rockets, to increase the Arrow's speed.

Later that afternoon, as Tsien wandered through the crowd, he felt a tap on his shoulder.

"Excuse me, sir." It was a young airman. "Are you Dr. H.S. Tsien?"

"Yes."

"Please come with me."

The airman led Tsien to the parking lot, out to a waiting black limousine with tinted windows. Tsien entered the car and, to his surprise, saw Air Marshal Campbell and Defence Minister Pearkes inside.

"Are you Dr. Tsien?" Campbell asked.

"Yes."

"Dr. Tsien," Minister Pearkes said, "the Russians have successfully launched an artificial satellite."

"A satellite!" Tsien exclaimed.

"It's called Sputnik," Campbell said. "TASS reports it was launched from Kazakhstan at 20:28 Moscow time."

"A satellite," Tsien repeated. "This is…incredible!"

"Yes, and now we need to totally rethink our strategies for the defence of North America," Campbell said. "Sputnik changes everything."

"Dr. Tsien, we have a job for you." Defence Minister Pearkes leaned forward. "The Government of Canada needs to learn everything you know about rockets."

* * *

Jeanne Sauvé, Minister of State for Science and Technology, yesterday announced the appointment of Tsien Hsue-shen, 64, as the first president of the newly created Canada National Space Administration (CNSA).

"I am honoured to be entrusted with the responsibility of leading the Canadian space program, and I will focus my efforts on ensuring that Canada remains at the frontier of science," said Tsien in a written statement.

The CNSA was created by an Act of Parliament that came into force on July 1, 1975. Established to develop the robotic arm that will be Canada's contribution to the NASA Space Shuttle program, the CNSA will also take over current space projects such as the Communications Technology Satellite and the Black Brant sounding rockets.

Born in Hangzhou, China, Tsien received a doctorate from the California Institute of Technology and played a key role in the early U.S. rocket programs before immigrating to Canada with his family in 1953. As a senior researcher at the National Research Council in the 1950s, Tsien worked on both the Velvet Glove missile and the Avro Arrow.

After becoming a Canadian citizen in 1957, he founded the Scientific Advisory Group at the Department of National Defence. Unlike many of his contemporaries, Tsien remained in Canada following the cancellation of the Arrow in 1959. Prior to his appointment to the CNSA presidency, Tsien was chair of the Department of Mechanical and Aerospace Engineering at Carleton University.

Tsien is married to Jiang Ying, the award-winning Chinese-Canadian soprano whose acclaimed final performance in the opera Eugene Onegin *opened the National Arts Centre in 1969. They have two adult children and recently celebrated the birth of their first grandchild.*

The appointment of Tsien is controversial. On CTV Question Period, Progressive Conservative industry critic George Hees questioned Tsien's suitability to head the CNSA. "The allegations of Communist affiliations that lead to his departure from the U.S. have never been unambiguously resolved," Hees said, citing also a 1970 CBC interview in which Tsien said the deaths attributed to Mao's disastrous Great Leap Forward were "probably exaggerated".

Sauvé defended Tsien's appointment, dismissing Opposition concerns as "baseless and irrelevant".

—The Globe and Mail, 16 July 1975

Afterword

To begin, a few words about a name. Hsue-Shen Tsien is the romanisation of his Chinese name 钱学森 under the older Wade-Giles system with the family name second in the Western custom. Today, his name is usually romanised as Qián Xuésēn under the Hanyu Pinyin system with the family name first in the Chinese custom. Since "The Son of Heaven" takes place from 1935 to 1975, I decided to use the Wade-Giles romanisation that would have been more common during that period. For example, the degrees he earned at American universities used the Wade-Giles romanisation of his name.

The opening epigraph of "The Son of Heaven" is taken from the 7 January 2008 issue of *Aviation Week and Space Technology* which recognised Tsien as its Person of the Year for 2007. In that year, China's space programme made two significant achievements: the launch of the Cháng'é Yīhào lunar orbiter and a controversial test of an anti-satellite (ASAT) weapon. While Tsien had no direct involvement with either of these activities (by then he was bedridden and would die in the following year, at the age of 97), the editors of *Aviation Week* justified their decision on the basis that Tsien had set the events in motion more than four decades earlier as the father of the modern Chinese space and ballistic missile programmes.

Much of the research for "The Son of Heaven" was sourced from *Thread of the Silkworm*, the definitive English-language biography of Tsien written by Chinese-American journalist Iris Chang.[1] The point of historical divergence in the story is subtle. Tsien's doctoral supervisor at the California Institute of Technology was the legendary Hungarian-American aerospace engineer and physicist Theodore von Kármán. Tsien earned his PhD in 1939 and stayed at Caltech as a professor, while von Kármán left in 1944 after recovering from surgery for intestinal cancer and eventually took a position to lead the Scientific Advisory Group at the Pentagon in Washington. When Tsien was first accused of being a Communist in 1950, he immediately declared his intention to return to China, an ill-advised move that brought further suspicion upon him. In the alternate timeline of "The Son of Heaven", von Kármán never fully recovers from the cancer surgery, staying at Caltech to advise Tsien and becoming a powerful advocate for his former student.

The real Tsien was deported back to China (not Canada!) in 1955. The Communist government quickly enlisted him to train aerospace engineers and develop a rocket program literally from scratch. The efforts of his team during the 1960s, with assistance from the Soviet Union, lead to the development of the Dōngfēng ballistic missile and the Long March (Chángzhēng) launch vehicle. Tsien also led the development of China's first satellite Dōngfānghóng Yīhào, which was launched in 1970 by one of the early Long March rockets. Perhaps embittered by his treatment at the hands of the U.S. Government, Tsien eventually became the very thing of which he had been accused: a Communist, at one point providing pseudoscientific cover for Máo Zédōng's disastrous Great Leap Forward.

When Tsien, his wife, and his two American-born children boarded the ship *President Cleveland* in September 1955, he told reporters that he would never set foot in the United States again. Tsien kept that promise, but decades later he would have one final and lasting influence on the U.S. space programme.

In 1988, the Reagan Administration initiated a policy that eventually permitted the launch of U.S.-built satellites aboard Chinese Long March rockets. Ten years later, allegations surfaced that sensitive information had been transferred to China during the investigation of three Long March

---

[1] Chang, Iris. *Thread of the Silkworm*, Basic Books, 1996.

launch failures in 1995 and 1996 that had been carrying U.S.-built satellites.[2] A committee of the U.S. House of Representatives chaired by Republican Christopher Cox of California investigated these allegations and issued a report in 1999 which concluded that China "has stolen or otherwise illegally obtained U.S. missile and space technology that improves the PRC's [People's Republic of China] military and intelligence capabilities."[3] Within the Cox Report was an entire section entitled "The Role of Qián Xuésēn in the Development of the PRC's Ballistic Missile and Space Programmes".

While the Cox Report also cited Iris Chang's book as one of its sources, unlike "The Son of Heaven" it did not claim to be a work of fiction. The Cox Report reiterated the accusations against Tsien in the 1950s and concluded that "the allegations that he was spying for the PRC are presumed to be true,"[4] a passive voice statement made without attribution (i.e. "presumed" to be true by whom?). In a time travelling twist worthy of science fiction, the Cox Report implied that Tsien had stolen secrets from the U.S. Titan intercontinental ballistic missile program, even though the request for proposal for the Titan missile was not issued until May 1955 and a prime contractor was only selected to start the project in September 1955—the same month Tsien was deported.

American space historian Dwayne Day has characterised the Cox Report as "a sloppy piece of work, full of extraordinary leaps of logic, dubious allegations, and incorrect facts."[5] Iris Chang herself had criticised the Cox Report for misusing her research, pointing out that "the U.S. never officially charged Tsien with espionage…[and] in the end they found no convincing evidence that he was either a Communist or a spy."[6] In December 1999, a group of professors from the Centre for International Security and Cooperation at Stanford University released an exhaustive 104-page study with itemized rebuttals of virtually all the claims made in the Cox Report, stating that the latter presented "a highly distorted and poorly researched picture"[7] that

2 Choi, Eric and Niculescu, Sorin. "The Impact of U.S. Export Controls on the Canadian Space Industry", *Space Policy*, Volume 22, Issue 1, February 2006, Pages 29–34.

3 Cox, Christopher (R-CA, Chairman). Report of the Select Committee on U.S. National Security and Military/Commercial Concerns with the People's Republic of China (Volume 1), U.S. House of Representatives, 09-066, 25 May 1999 (Declassified Report), Page xii.

4 Cox, Pages 177–179.

5 Day, Dwayne A. "A Dragon in Winter", TheSpaceReview.com, 14 January 2008, https://thespacereview.com/article/1035/1.

6 Ni, Perla. "Author Denounces Cox Report: Iris Chang Tells Conventioneers That Her Research was Misused", *Asiaweek*, 3 June 1999.

7 May, Michael (editor). "The Cox Committee Report: An Assessment", Centre for International Security and Cooperation, Stanford University, Dec. 1999, Page 22. https://carnegieendowment.org/pdf/npp/coxfinal3.pdf.

was riddled with "factual errors, misused evidence, incorrect, misleading, or non-existent citations, and implausible interpretative spins of existing research."[8]

Nevertheless, the Cox Report had a profound and enduring impact on U.S. space policy. Section 1513 of H.R.3616, the Strom Thurmond National Defence Authorization Act for Fiscal Year 1999 (named for a segregationist former senator) transferred responsibility for the export control of space technology to a restrictive regulatory regime under the U.S. State Department known as the International Traffic in Arms Regulations (ITAR).[9] Under its original implementation, ITAR made no distinction between civilian and military space technologies, nor did it differentiate between adversaries like China and nominal allies such as Europe and Canada. ITAR imposes significant challenges on international collaboration with the United States on civil, commercial and scientific space projects, and it is extremely difficult for non-American citizens to find employment in the U.S. space industry. In 2011, the U.S. Congress passed a law that included an amendment introduced by Republican Frank Wolf of Virginia that prohibits NASA from engaging in direct bilateral cooperation with China on space projects. Despite some reforms, both ITAR and the Wolf Amendment remain fundamentally in effect to this day.

To conclude, a few words about names. Hanssen and Roberts, the two fictional FBI agents who first question Tsien in "The Son of Heaven", are named for Robert Hanssen, the treasonous American-born FBI agent who spied for the Soviet Union and then Russia against the United States for more than twenty years until he was finally apprehended in 2001. He is currently in a maximum security prison serving fifteen consecutive life sentences without the possibility of parole.

---

[8] May, Page 37.
[9] Choi and Niculescu, Page 30.

From a Stone

Stone.

Cold, hard, unimaginably ancient stone. A rough, irregular aggregate of it, angular, cratered, pockmarked, and pitted. The stone was dark, with an albedo of less than five percent. It tumbled lazily about its major axis with a rotation period of 19 hours, 53 minutes. A third of it was frozen in shadow, while the rest roasted in the incessant light of the Sun. It was only ten by five by four kilometres in major dimension—on the scale of the Solar System, practically microscopic.

But the stone was not too small to escape human curiosity.

"*Hold at three hundred metres for IPS parameter update.*" The voice of pilot Ben Dixon came over Pierre Caillou's spacesuit radio. Dixon and the mission commander, Poornima Bhupal, were monitoring the EVA from the bridge of the UNSDA spacecraft *Harrison Schmitt*.

The astronaut-geologist entered the command into his manned manoeuvring unit. "OK, Ben," Pierre replied. "Hold at three hundred metres." He glanced to his left and right. Diane Sokolowski and Marvin Shipley, the other geologists from the *Schmitt*, were flying in formation with him, both piloting their own MMU thruster packs.

The asteroid's bulk loomed ever larger as the astronauts approached. Pierre kept an eye on his lidar rangefinder. At that precise moment a nitrogen jet fired, putting his MMU in an attitude-hold mode three hundred metres from the surface.

"I am holding at three hundred metres," Pierre said. His colleagues reported the same.

© The Author, under license to Springer Nature Switzerland AG 1996
E. Choi, *Just Like Being There*, Science and Fiction,
https://doi.org/10.1007/978-3-030-91605-3_4

"*Copy*," Ben replied. "*Transmission of IPS updates will commence in five seconds…Mark! Note the new basis vectors are for an asteroid-based frame of reference, with the origin located at the geographic centre of 2046-PK.*"

"Understood." A green light on Pierre's inertial positioning system blinked. "My IPS reports updated state vectors received and installed." Diane and Marvin reported their navigation units also were ready.

Ben enunciated his next words formally. "*To EVA crew, I have the following messages from the CAPCOM and Commander Bhupal: Mission Control Darmstadt is 'go', Schmitt is 'go'. You are cleared for final approach to 2046-PK. This is the last hurrah, people. Make it good.*"

"Understood." Pierre's aft thrusters fired a short burst, putting him in motion once more. The dark surface of the asteroid rushed up to him. Pierre executed a pitch-back manoeuvre to put his feet "down" before the negative y-axis thrusters came on to slow his touchdown. "…20 metres…10…5…1…Contact!"

"I have contact also," said Diane.

"I'm down!" followed Marvin.

Pierre surveyed the magnificently desolate scene about him. "They say you can't draw blood from a stone –"

"But you can always get knowledge!" Diane's exuberant voice cut in. They would later swear it was not rehearsed.

The astronauts went about gathering samples, hopping across the asteroid like grasshoppers. Pierre presently found himself near one of 2046-PK's larger craters. He decided to sample the rim material, and his MMU dutifully delivered him to the edge. As he readied his tools and glanced at the bottom of the crater, he noticed something unusual.

The Sun was shining at an angle across the impact, producing a semicircular shadow that bisected the bottom of the bowl. At the top of the arc, a second, smaller semicircle jutted out from the shadow. There appeared to be another crater at the bottom of the larger, but somehow it didn't look right.

"Pierre for Diane."

"Go ahead."

"Have a look at this." He transmitted his camera video.

"That's interesting," Diane replied after watching Pierre's feed on her multifunction visor. "I'm coming to have a look."

Moments later, she landed beside him.

"So, what do you think?" Pierre asked.

"Looks like another crater."

"That's what I thought. But look at the edge. It's not very circular. In fact, it's pretty ragged."

"Must be really old," Diane concluded. "A few billion years of dust and micrometeoroid impact will do that."

"Of course. But look at the larger crater." Pierre's hand swept out a curve. "The rim is *smooth*. That smaller crater is on top of this larger one, so the smaller one must be more recent. So, how could it have experienced more degradation than the older, larger impact?"

"I don't know," Diane said.

"Pierre for Ben. Are you seeing this?"

"*Affirmative,*" Ben replied from the distant *Schmitt*. "*Poornima and I concur with your assessment, but we don't have any new ideas.*"

"We should have a look," Diane suggested.

There was a pause as Ben consulted the commander. "*All right. We'll let Darmstadt Mission Control know of the change. But don't be too long. Time is not on our side, and we're behind schedule already.*"

"Thanks, Ben." Pierre turned to Diane. "All right. Let's go." Puffs of nitrogen sent them skyward from their perch on the rim to a point above the inner crater.

"It doesn't look round," Diane observed.

"No, it doesn't." Pierre activated his lidar rangefinder. "I am getting a uniform depth sounding. It is not a circular depression."

"*Ben for Diane and Pierre.*"

"Yes, what is it?" Diane sounded irritated.

"*Uh, sorry to bother you, but Darmstadt wants to inform you the back room boys are getting restless.*" The "boys" were the geologists monitoring the expedition from their own room at Mission Control in Darmstadt, Germany. EVA schedules were as tightly scripted as Shakespeare, and the three-day time line for the survey of 2046-PK was particularly tight. The message was a subtle hint for Pierre and Diane to get back on track.

Diane gave Pierre the hand signal to switch to a private channel. "Oh, for crying out loud. We're not automatons. If they wanted robots, they should have damn well sent them!"

Pierre switched back to the common loop. "Ben, we were scheduled to sample the bottoms of a few craters tomorrow. There's no harm in getting started today. Besides, we seem to have something rather unusual here."

"*Yeah, I can see that.*" A pause. "*OK, go ahead. Poornima says she'll try to smooth any ruffled feathers back home.*"

"Thanks," Diane said.

They descended until they hovered right over the opening.

"Look at the edges!" Diane pointed. "It's definitely not circular. It looks kind of like a rounded-off pentagon or something."

"Well, this is certainly not a crater," Pierre said. "I am reading a depth of four hundred and fifty metres. This is a tunnel." He switched channels. "Pierre for Marvin."

"Go ahead."

"Take a look at this." Pierre sent him the video.

The Canadian astronaut sounded puzzled. "I don't know. Could be a lava tube. But on a rock this small? Beats me, Pierre. I'm just a prairie boy from Manawaka."

Pierre played a light down the hole, but the darkness swallowed it. "I want to go to the bottom. Ben?"

"*Pierre* –"

"Ben," Diane cut in. "Tell Poornima to tell the back room boys to cool it. We've got something really worth…ahem, looking into here. Look, just let him have a quick look-see. OK?"

Ben sighed. "*Very well. You stay outside to relay Pierre's signal. Be careful, OK?*"

"That goes without saying," Pierre replied dryly. He started his MMU and plunged into the unknown. His suit lamp produced a small circle of illumination that tracked down the wall as he descended. It was pitted and scarred from billions of years of micrometeoroid and dust impact, but the surface did not seem as coarse as the outside of the asteroid.

Pierre stopped. "One hundred metres down. I'm going to take a sample of the wall."

Another sample was taken at the two hundred metre mark. By now, Pierre was certain there was something strange about the tunnel. "There are fewer signs of geologic modification now. There is less evidence of erosion and wear." He touched the rock. "In fact, it feels almost…a little *flat*."

"*Uh, copy,*" Ben replied.

Pierre resumed his descent. As the wall scrolled by, it became apparent that it was not only becoming flatter, but the surface was smoothing out as well. It was like a beach, where the rough sand trampled by innumerable feet merged into the smooth area washed by the waves.

He stopped at three metres for another sample. Pierre ran his hand along the wall. With the exception of a few dents created by micrometeoroids that had managed to get this far, the wall felt uniform. He transmitted his video to Diane.

"What do the other surfaces look like?"

"Other surfaces?" The obvious hit him with a start. All this time, he had blithely assumed the tunnel was circular. But if he were staring at a flat surface, this clearly could not be the case. He rotated his MMU.

There was a corner—and another, and another. Six in total, formed by the intersection of the walls. "This tunnel…it is not circular! It is a hexagon!"

"How can that be?" Ben asked.

"That's incredible," Diane breathed. "But at least it explains one thing. This opening isn't really any older than the outside crater. It just looks like it had experienced more modification because it's really an eroded hexagon instead of a circle."

"I'm going straight to the bottom," Pierre announced as he started his MMU. The smooth walls scrolled by as the display on his rangefinder counted down. But just three metres from the bottom, his suit lamp seemed to black out. "What –" he began, but his feet were already touching bottom.

He rebounded off the floor before his MMU went into attitude-hold. The asteroid's feeble gravity made it impossible to really "stand", so Pierre just floated a few centimetres above the rock.

"I am at the bottom." He looked down at where his feet had touched, and to his surprise discovered his lamp had not malfunctioned. The light revealed a small circle of rock, marred only by a handful of impacts. Puzzled, he looked up again—and gasped.

"Mon Dieu!"

"*Pierre, what is it?*" Ben asked urgently. "*What the hell is that in front of you?*"

"There is another tunnel in front of me. The opening is hexagonal." He checked the lidar. "It is two hundred and fifty metres deep." He turned, and counted five other walls. Each one had an opening identical to the first.

"Oh, my God," Diane whispered.

"What's this?" Marvin barked. "What's going on? Pierre!"

"*Cut the chatter on the loop!*" Ben snapped. "*Pierre, what is your appraisal of this formation?*"

"Ben, I believe…" Pierre swallowed. "I have no explanation at this time as to how these features were produced. This sounds crazy, but Ben…I don't believe this is a natural phenomenon. This is something *alien.*"

* * *

The stone was discovered by astronomers at the Shapley Observatory on the lunar farside, and was given the designation 2046-PK after the year, half-month, and order of its discovery. Its highly elliptical orbit had a period of 136.9 years, and was inclined 18.3 degrees to the ecliptic. This seemed to suggest the object was a spent comet. But some astronomers believed 2046-PK was a visitor from beyond the Kuiper Belt, an extrasolar body captured by the Sun when it experienced just enough gravitational perturbation to close its hyperbolic trajectory into an ellipse.

The United Nations Space Development Agency proposed a mission to 2046-PK, but politics and funding conspired against it. Fortunately, the laws of celestial mechanics offered an alternative. One UNSDA spacecraft—the *Harrison Schmitt*—was already in solar orbit, finishing a survey of Apollo-class asteroids. Through a bit of orbital legerdemain, the ship was diverted to rendezvous with 2046-PK at the point it crossed the ecliptic on its ascending node, exiting the inner Solar System.

It was hardly an ideal situation. The *Schmitt* was already at the end of its nominal flight, and its crew had to ration their supplies of food and oxygen to support the extended mission to 2046-PK. Furthermore, the unfavourable rendezvous geometry consumed so much fuel the astronauts were left with just three days to study the asteroid. After three days, the burn window would close, and the *Schmitt* would no longer be able to make an Earth-return trajectory.

But a hit-and-run mission was better than none. The astronomers and geologists were used to the compromises inherent in "government science". They consoled themselves in the belief that the only unique thing about 2046-PK was its orbit, and that it probably wasn't much different than the dozens of asteroids the *Schmitt* had already surveyed.

Then Pierre Caillou discovered the Beehive.

* * *

The flight controllers in Darmstadt were in a conundrum. They were, after all, trained to deal with mechanical failures and schedule problems, not first contact. Despite the looming departure deadline, Darmstadt cautiously decided to forbid further exploration until the appropriate authorities were consulted. The EVA astronauts were instructed to return to the *Schmitt* until new orders were issued.

While awaiting their arrival, Commander Poornima Bhupal watched the video again. "It's like a beehive."

"Huh?"

"A beehive," she repeated.

"Yeah, I thought so too."

"What do you think..." She gestured at the screen. "What do you think this is? Who were they? What does it all mean?"

Ben shrugged. "How the hell do I know?" He turned to another monitor. "They're back. I'm opening the airlock." He left the bridge to help the geologists doff their spacesuits. When that was completed, the crew gathered on the bridge.

"What's the plan?" Pierre asked.

"We're waiting for instructions from Darmstadt," Bhupal said.

"When can we expect that?" asked Marvin.

"Should be any time now, I hope."

Moments later, Ben announced, "Incoming message."

Jason Ho, the CAPCOM on duty, appeared on the screen. "Commander Bhupal, crew of the *Schmitt*." He was obviously reading from a prepared statement. "On behalf of everyone here at Mission Control, Flight Director Pearson, the engineers, the scientists, geologists...indeed, on behalf of all the people of Earth, I congratulate you on this remarkable discovery. This is a truly momentous occasion. Human history has been altered..."

Diane silently mouthed "blah, blah, blah".

"...will never be the same. But precisely because the nature of your discovery—proof of extra-terrestrial life—is so extraordinary, it becomes necessary to take extraordinary precautions. We must act very carefully, for we do not know the consequences of any mistakes.

"As of now, 22:43 hours, the SETI Protocol is in effect. The Secretary General and the First Minister have been notified. A tiger team of engineers and scientists from all UNSDA centres is being assembled to recommend the next course of action. But any such action must first be approved by the Peace and Security Council. So, until the PSC makes policy, the crew of the *Schmitt* is ordered not to attempt another EVA. You will hold your position until further instructions, which you will receive shortly."

Jason stopped reading, and his expression changed. He now addressed his fellow astronauts, his friends. "Poornima, Ben, Diane, Marv, Pierre...hang in there, OK? We'll figure this out."

The screen went blank.

"Further instructions shortly. Right." Diane turned to Marvin. "By the way, what time is it?"

"22:50."

"No, I mean in New York."

"It's... what? Three or four in the morning, I think." Pierre saw what Diane was getting at. "Oh, no."

"'Oh, no' is right. They're all in bed. It'll take hours just to wake them up, get them their coffee and doughnuts, put them in a room...Then they'll yak for hours more." Diane clenched her fists. "Further instructions shortly.

Right. There's classic government science for you. We'll be lucky if we're left with even a day to explore!"

"Well," Pierre said, "while we're waiting, perhaps we could analyse –"

"You will do nothing of the sort," Bhupal interjected. "I want the three of you to get some rest." She emphasized the last three words with jabs of her finger. "When the powers-that-be finally decide our next move, I want all of you to be refreshed and ready to go as soon as the word is given. We have only two more days here and we must make the most of them. Understood?"

"Sure," Marvin said. Pierre nodded, and Diane murmured in agreement.

"All right. Grab a bite to eat, maybe. But then you hit the sack."

The geologists gathered in the *Schmitt*'s galley. Pierre and Marvin both prepared meals, but their appetites were wholly divergent. Marvin attacked his food with the vigour of a starving person, while Pierre poked and prodded at his tray, eating very little.

Diane's eyes wandered back and forth between them. "How can you guys think of food at a time like this?"

"Actually, I'm not hungry." Pierre put down his fork.

Marvin swallowed. "Hey, what's wrong? I'm within my ration."

"I just think it's a little... weird." Diane shrugged. "I mean, here we are. We've just discovered proof of intelligent extra-terrestrial life, and what's the first thing we do? Graze."

"Wait a minute." Marvin wiped his mouth. "Did you just say 'intelligent life'?"

"Yeah."

"Uh, sorry Di. I have to disagree with you."

She smiled. "Oh?"

"I know what I saw," Pierre said. "You've all seen the video. Those caverns were definitely not natural."

"Oh, I'm not disagreeing with that," said Marvin. "They're alien, no doubt about it. But whether the builders—the aliens themselves—whether they were intelligent, now that remains to be seen."

"But the shape, those walls..." Diane shook her head. "How could something build something like the Beehive and not be intelligent?"

"Beehive?"

"Beehive asteroid," Diane explained. "That's what the media's calling 2046-PK."

"Beehive..." Marvin mused. "Yeah, that's good. See, you don't have to be smart to build complicated things. Just like bees. Sure, they build complex hexagonal hives, but they're as dumb as posts. On the other hand, whales don't build anything, and they're probably smarter than I am."

"That's a good point, Marvin," Pierre said. "But it raises another question. What would you consider as proof of intelligence?"

"Math," Diane replied immediately.

Pierre nodded. "OK. Perhaps even music then, since it is very much a mathematical construction."

"Like whale song," Marvin suggested.

"Exactly," said Pierre. "Something to do with math. In fact, one of the mathematicians—I think it was Gauss—once suggested that a giant representation of Pygathoras' Theorem be put in the Alps in hopes of signalling any Martians."

Marvin shrugged. "Well, I hope you two aren't too disappointed, but that asteroid isn't exactly spouting calculus and Chopin."

Commander Bhupal entered the galley. "What are you still doing here? Talking? The people on Earth are talking. New York is talking. You should not be talking. You should be resting, so when they stop talking, you'll be ready to go."

The geologists laughed. "I'm off," Marvin announced as he left the galley. Diane did the same, throwing up her hands in mock surrender.

Pierre retired to his cubicle. He told the lights to dim, climbed into his sack and shut his eyes. Sleep came almost immediately.

He dreamed.

He saw himself floating through one of those hexagonal tunnels. It was pitch dark, with only a small circle of light from his lamp to show the way. The tunnel took a turn, and he pushed himself off a wall to change direction as he had so many times before in the passageways of spacecraft. But this was no spacecraft—at least, not a human one.

The tunnel abruptly ended, and he found himself in a large cavern. He played his light along the walls and discovered the openings of four other tunnels. Where he had expected a fifth, there was something else.

It was a humanoid figure.

He approached. The being was imbedded in the rock itself, like a relief sculpture carved into the wall. It was about his height and build, dressed in a garment that resembled some kind of spacesuit. He tugged at the figure, but it could not be dislodged.

The creature wore a helmet, the front of which was shielded by a unit like a multifunction visor. Pierre put out his hand and raised it.

There was a face.

It was his own.

* * *

Fourteen hours after the initial discovery, the PSC reached a decision. The relatively prompt action of the Council was, even Diane admitted, something of a miracle. Nevertheless, there were now just two more days before the *Schmitt*'s window for its vital trans-Earth injection burn closed.

After eating his breakfast ration, Pierre joined the crew on the bridge to hear the outcome of the PSC's emergency session. On a monitor, CAPCOM Oleg Solovyov read the communiqué.

"The Peace and Security Council of the United Nations has reached a decision on Resolution 2071–57, sponsored by the United Nations Space Development Agency, requesting the further exploration of the extra-terrestrial features found inside asteroid 2046-PK. The result of the vote was as follows…"

"Oh, forget the breakdown," Diane groaned.

"Hey," Marvin hushed her. "I'm interested."

"In favour: China, the European Union, Russia, Israel-Palestine, India, Japan. Opposed: Brazil –"

Cheers erupted on the bridge.

"– and the Unified African Republics. Canada and the United States abstained. Resolution 2071–57 was thereby approved."

"Here come the provisos," Ben cautioned.

"Further exploration of asteroid 2046-PK is authorized, subject to the following conditions and precautions: During EVA, Pilot Benjamin Dixon shall be stationed in the airlock, fully suited, ready to exit on short notice. Commander Bhupal shall act as the bridge intravehicular crew member in his place. All EVA crew, as well as Pilot Dixon, shall carry a heavy-duty laserdrill on their person at all times while outside the spacecraft.

"Activities on the asteroid itself shall be subject to the following conditions and precautions: One EVA crew member shall station themselves at the end of the entrance tunnel. The remaining crew members may then enter one of the secondary tunnels, but both must enter the same tunnel and stay together at all times.

"The contingency EVA tethers shall be used for communications by crew inside the asteroid. Under no circumstances are EVA crew to exceed the range limits imposed by the tethers. They are to gather as much information as possible while exercising extreme caution. This is humanity's first contact with extra-terrestrial intelligence. The importance of your mission cannot be overstated. Further information and instructions will be uplinked as required. This transmission ends…now."

"Extra-terrestrial *intelligence*…" Marvin shook his head.

"These so-called precautions," Diane sneered. "They're silly."

"Perhaps they are a little conservative," Pierre agreed. "But for the time being, we 'shall' follow them."

"All right, people!" exclaimed Commander Bhupal. "We haven't much time. Let's get moving!"

Forty-five minutes later, Diane, Marvin, and Pierre returned to the rocky surface of the asteroid. A communications relay was stationed at the mouth of the entrance tunnel. Pierre and Diane descended to the end of the shaft first, while Marvin hitched his tether to the comm unit before following.

"Comm check on relay," Marvin said.

"*Good comm check,*" Bhupal replied. "*Please proceed.*"

The astronauts doffed their MMUs and anchored them to the walls. Diane then tethered herself to Marvin. Pierre would remain a free-floater. After a final check of their suits and equipment, Marvin announced, "OK Poornima, we're ready to go."

"*Copy. Be careful...and good luck.*"

Marvin surveyed the secondary entrances. "Well, take your pick."

"Enie, meanie, minie, moe..." Diane recited. Nervous chuckles echoed over the loop.

"Let's take this one," Pierre suggested, indicating the opening he happened to be closest to.

"As good as any," Diane said. "Let's go."

It was a historic moment, and Pierre wished he and Diane could come up with another memorable phrase. They could not, so he said simply, "We are going in."

"Good luck," said Marvin.

"*Darmstadt—and I—also wish you good luck,*" Commander Bhupal said. "*And...be careful.*"

Pierre went in first, followed by Diane trailing the thin, threadlike communications tether.

"The walls are similar to those observed in the outer entrance tunnel," Pierre reported. "They are hexagonal. They are also quite smooth."

Diane stopped. "Pierre, hold on a sec."

"Yes?"

"Let's quantify that statement. Let's see how smooth this really is." She produced a gauge and ran it along a wall. "I've got a five millimetre deviation over a thirty centimetre length." She ran it along another wall. "Six over thirty here."

"Measure the angle between the walls," Pierre suggested.

"OK, the angle between the walls is...fairly close to thirty degrees," Diane reported. "Not exactly." She pointed. "This one is twenty-seven point five degrees, while that one is twenty-eight point nine. There's an error of plus or minus three degrees."

"Marvin for Diane. You sound a little disappointed."

"Doesn't matter, Marv. These walls were machined. They had to have been, and you need smarts to run a machine."

"*Poornima for EVA crew. Please leave your speculations for later. Pierre, Darmstadt requests more samples.*"

The back room boys again.

"Right, Poornima. Thank you. I was just going to do that." Pierre put aside the PSC-mandated laserdrill and produced a hammer and chisel. He brought them up to the wall and was about to strike, when he paused. Even if Marvin was right, that the beings that dug these tunnels had the brains of a maggot, it didn't diminish the fact that this was an alien artefact. Pierre felt a little bit like a vandal.

But he took the sample anyway.

They continued. The tunnel took a turn, and Pierre experienced a moment of deja vu as he and Diane pushed off a wall to match the curve. The feeling intensified when they emerged from the tube.

"We're in some kind of chamber," Pierre reported. "It is roughly spherical, but the walls are not finished as they were in the passageways." He panned the camera. "There are entrances to other tunnels. However, they are not distributed in any regular pattern." Two were fairly close together, while the rest were scattered in random positions.

"Poornima for EVA crew. It's time to call it a day. Diane and Pierre, finish what you're doing and start heading back."

Pierre checked his oxygen gauge. "OK, Poornima. Understand it's a wrap." He took a final sample, then panned the camera about one last time. "All right Diane, time to go home."

"Uh, OK."

Pierre started back, but Diane did not follow immediately.

"Diane?" He switched to a private channel. "Is there something you want to say?"

A pause. "No."

Two hours later, the geologists were back aboard the *Schmitt*. After a quick supper of ration packs, they went to the lab to do a preliminary analysis of their data.

"You know," Marvin began, "this rock would be a bundle of mysteries even without the alien thing."

Diane feigned innocence. "Why are you looking at me?"

Pierre asked, "You are certain about the age?"

"No question." Marvin consulted his tablet. "Two point eight billion years."

"Younger than the Solar System," Pierre mused.

"Meaning it did come from outside the Solar System," Diane concluded.

"Maybe," Marvin said. "There's something else." He told the computer to project a holographic image of the asteroid above the table. The potato-like rock floated whole before their eyes for a moment before the computer sliced it in half, revealing Marvin's model of the internal structure in cross-section.

"I based this on the samples you took inside the caverns and the surface measurements made by the remotes. The surface is composed of a thin veneer of achondrites, while the bulk of the asteroid itself is composed of metamorphosed chondrites." The colour scale changed. "However, as we go deeper, we find that the percentage of siderophile elements is increasing –"

"Iron," Pierre interrupted. "Stony-iron materials."

"Yes," said Marvin.

Pierre bridged his fingers. "How much?"

"From the surface to the deepest point you and Diane sampled, there's an increase of nine point one percent, within an error of –"

"That much!" Diane exclaimed. "But we didn't go that far down!"

"The asteroid is differentiated," Pierre concluded. "It has physically and chemically stratified zones. But how? It's so small."

Marvin shrugged. "It could have been heated to a molten state sometime in the past. But you're right, its gravity should have been too weak to sift the elements as it came back together. If it did, though, it would have reset the isotopes. It could be a lot older than we measured. Maybe it was formed in the Solar System after all."

"But what could have provided the heat?" Diane asked.

"Planetessimal bombardment, maybe?" Marvin suggested.

"Could be." Pierre pointed. "But look at how young it looks. There are very few large impacts."

"What about radiogenic heating?" asked Diane.

Marvin shook his head. "No, this rock's cold. No sign of radioactive decay. And the magnetic field is only a few hundredths of a gauss."

Diane smiled. "You know Marv, I think you're right. This rock would be just as interesting without those alien tunnels." She became serious. "But they're there, and tomorrow's the last chance we'll get to study them."

She erased the cross-sectional profile, restored the asteroid to its true shape, made the image semi-transparent, and traced out a network of nodes and lines. "This is where we've been. We've only explored seven percent of those caverns."

"That percentage isn't likely to get much bigger after tomorrow," Pierre observed.

"Damn tethers," Diane muttered. "The fact is, if we follow the pattern we did today, I doubt we'll cover much more ground or learn anything new."

"What do you suggest?" Marvin asked.

"I suggest we try to go to the centre of the asteroid."

Pierre stiffened.

"Di," Marvin raised his hand like a student, "may I state the obvious? There's no way you two are going to get near the core. We don't even know if the tunnels will take you there. Assuming they do, the tethers are only—Now *hold on* there…"

"We might be able to reach it if we tethered in series, going in single file."

"But we don't know how those passages twist and turn," Marvin objected. "Even strung together, you probably wouldn't reach the core. Besides, the PSC guidelines clearly state that the two of you must stay together while —"

"Oh, screw them!" Diane spat. "Classic government science. They're not even here. We've only got one more EVA and we've got to make it count. I mean, let's suppose—Pierre?" She snapped her fingers. "Hey, Pierre!"

"Huh?"

"Are you all right?"

"I was just thinking."

"About what?"

"Well…" Pierre shrugged. "This is kind of silly, but last night I had this weird dream." He described it.

"That *is* weird," Diane breathed.

"Do you think it means anything?" asked Marvin.

Pierre shrugged again. "It was probably an anxiety dream. I have those sometimes."

"You're probably right," Diane said. "Anyway, Freud can wait. We've got bigger things to worry about." She looked at her colleagues. "It's time we fish or cut bait. Marvin?"

"Look…" Marvin traced a finger along the table. "It's not that I'm not curious. I am. But this isn't some movie where the rebellious officers break the rules, save the day, and end up getting promoted. It won't matter *what* we find down there: iron, Pierre's evil twin brother, even King Tut's tomb. Our careers will be *finished*. We won't be punished—well, not publicly anyway. We'll be honourably retired, maybe even given a desk job if we're *really* lucky, but we'll never fly again. And that's if we *find* something. If we don't…"

"We'll find something," Diane declared.

"*That* worries me, too," Marvin said. "Diane, we have no idea what could be down there!"

"Exactly!" she exclaimed. "If we knew, we wouldn't need to go. The point of going is to find out!"

"Why don't we ask Darmstadt first?" suggested Marvin.

"And they'll take it to New York. Another PSC meeting. More debate, more delay. By the time we *get* approval—assuming they say yes—it will be time to go."

"Can the mission be extended again?"

Diane shook her head. "Impossible. The TEI burn window closes after tomorrow. If we don't leave then, we're never leaving." She paused. "Marvin, please. Are you in or not?"

Marvin took a deep breath. "Yeah, count me in."

"Pierre?"

"Yes, let's do it." He tried to smile. "Tomorrow, we will solve all the mysteries of the Beehive asteroid, right?"

Pierre did not dream that night.

* * *

The geologists descended to the bottom of the entrance tunnel for the last time. As Pierre checked his tools, Marvin slipped Diane the extra tether spool he had smuggled out. They could have used another, but taking more might have been noticed. Since the tethers were their only link with the *Schmitt*, as long as the telemetry stream was uninterrupted there should be no suspicions until they were beyond recall.

"Are the IPS units still on the asteroid-centred coordinate system?" Diane asked.

"Yes." Pierre switched to a private channel. "This tunnel took us deepest yesterday."

"We'll start there, then." Diane switched back to the open loop. "I'm ready."

Pierre signalled the *Schmitt*. "OK, Poornima. We're ready to go in."

"*Copy that, Pierre. Darmstadt says 'go', and I concur. Be careful, and good luck.*"

"We'll need that today," Pierre muttered. Louder, he said, "OK, Marvin. We'll see you when we get back."

"Good luck."

Pierre and Diane floated into the tunnel. They stopped a hundred metres in and looked back the way they came. Diane turned the camera away and keyed a private channel. "Come on in, Marv. The water's fine."

"I'm on my way." Moments later, he was reunited with his colleagues.

The three continued in single file with Pierre in the lead, followed by Diane, with Marvin taking up the rear. They entered a cavern. Like the others before, this one presented them with five other entrances to choose from.

"Where to now?" Diane asked.

Pierre fired his lidar down each tunnel and entered the reading into his IPS. "That one."

They drifted together in the darkness for an hour before being forced to a halt. Marvin had reached the end of his tether.

"That's it for me."

"Thank you, Marvin," Diane said.

"Are you going to be all right?" Pierre asked.

Marvin's multifunction visor was in transparent mode, so his smile was visible. "Sure. Hey look, if anything comes after me..." he hefted his laserdrill, "I'll poke their eyes out with this!"

Now Diane smiled. "They might not *have* eyes."

"Then..." Marvin lowered his voice, "I'll stick it up their ass!"

Pierre and Diane pressed on. They probed the tunnels and caverns like a computer working a recursive search tree. The pair selected the likeliest branch and went as far as they could. If the path did not take them closer to the core, they returned to the last cavern and selected another. The two had backtracked, turned, retraced, and doubled back and forth so many times Pierre's internal sense of direction was completely gone. If not for the tethers and the IPS, they would have been hopelessly lost.

Diane's tether gave out while they were in a cavern. "Oh, damn."

"Well, this was hardly unexpected. Guess I'm on my own now."

She started unlocking her spool.

"No, don't! I'm going on myself. We agreed, remember?"

"Yeah." Diane let go of the device. "Say hello to your evil twin brother for me?"

"Of course."

She hugged him. "Be careful."

"I will."

Pierre released himself from the embrace. He and Diane then switched off their video feeds to the *Schmitt*.

"*Poornima for EVA. What's going on?*"

"Uh...our cameras are dead," Diane stammered.

"Both *of you?*" The commander sounded incredulous. "*How can –*" The voice link was also cut to avoid answering any more embarrassing questions.

"Good luck."

Pierre nodded. He entered a tunnel, and moments later emerged in another cavern. Again, there were five other entrances. Pierre probed each with the lidar, then checked the IPS and the status of his tether.

"Pierre for Diane. I'm only seven hundred metres from the geographic centre of the asteroid, but I don't know if I'll make it. I won't if these tunnels don't take me there directly."

"Do what you can, but get a move on, OK? Poornima's probably sent Ben out by now to find out what the hell we're doing."

"Right."

He was about to set off when he stopped. Something seemed *different* about this cavern. He panned his camera about to document the scene before diving into another tunnel.

His eyes scanned the walls of the passageway as they rushed by, and this time he was sure there was something different. He paused to run a gauge along the surface. There was only a one millimetre deviation over a thirty centimetre run—a far better finish than the outer tunnels. The angles between the walls were also checked. They were all thirty degrees, to within one-tenth of a degree.

Suddenly, his suit computer came on. *"Warning. Check tether status."* A gentle tug on the line confirmed the increase in resistance.

"Pierre for Diane. I have reached the end of my tether."

"How far are you from the core?"

"Only three hundred metres." Pierre made up his mind in an instant. "I'm going to disengage the tether."

"That's –" The line was disconnected.

Pierre turned, pushed off a wall, and darted down the dark passage. He moved quickly without the tether. A sharp corner surprised him, but he managed to deflect his trajectory off the side without colliding.

He emerged in a cavern. His momentum carried him across the void to a far corner where two walls met. Pierre stopped himself and consulted his IPS. The reading was in single digits. He was at the core of the Beehive.

The geologist panned the camera, and when his mind assembled a complete picture from the fragmentary glimpses provided by the light, he gasped.

This place was completely different from the others. Those had been just crude hollows, without definite form or pattern. This one was, for want of a better word, a *room*. It had six equally sized walls which formed a perfect hexagon. Each wall in turn had a hexagonal tunnel opening centred on it. But curiously, the room did not have three-dimensional symmetry. There was a definite up/down orientation, a paradoxical arrangement in a weightless environment.

His heart beat faster, and he noticed he was trembling. He forced himself to stop and take a few deep breaths. Slowly, he played his light along the "ceiling". As the circle of illumination moved, more of it came into view. Then, the light touched something, and he saw—a figure.

Not a humanoid—a geometric figure. It was a six-sided solid, but not a perfect hexagon. Its length was twice its width. Upon closer examination, he could see there were actually two shapes side by side. The one on his left was a full hexagon, consisting of dark rock with a protrusion in the centre. The area on the right was covered with lighter material. He picked at it with a scalpel. Grains of regolith floated away in chunks.

Pierre scraped along the edge of the hexagon, continuing downward until he hit the bottom edge where it met the surrounding wall. He found a groove running along the perimeter. The lip of rock was holding the hexagonal slab against the wall on four sides, while the packed soil was keeping it in place from the other two.

Pierre scraped at the dirt. When most of it was cleared, he drove an anchor into the wall and grabbed it with one hand while holding the protrusion on the hexagon with the other. He pulled. It jerked and stuck, but it had moved. Taking out the pick, he cleaned out more before trying again.

Slowly, the hatch opened.

Inside the cavity were…stones.

Just stones. Six of them, imbedded in the wall, arranged in a circle with size increasing clockwise.

A set of stones.

"What the hell is this?"

A bed of regolith cradled each of them. Whoever—whatever—had done this had carved out sockets in the cavity, placed the stones in them, and sealed them in place with the mortar-like soil.

Pierre Caillou used a fine pick to pry the orbs free, and gently placed each of them in a sample bag.

* * *

In all the time he had known Commander Bhupal, Pierre observed her to respond to anger in one of two ways. She would either shrug her shoulders, or go ballistic. Upon their return to the *Schmitt*, the first thing the geologists did was report what they had done, and why they had done it. Pierre described his discovery at the core of the asteroid, and Diane asked rhetorically whether it had been worth the risk.

Poornima Bhupal shrugged her shoulders.

The geologists gathered in the *Schmitt*'s lab. Their futures were uncertain, but for now, their minds were much too busy pondering other things.

"From the samples I took in that room," Pierre began, "I have confirmed the core is stony-iron. The age of the asteroid is also verified as 2.8-billion years."

"I still don't see how it could be differentiated," Marvin said. "The asteroid's just too small."

Pierre shook his head. There were so many unanswered questions. "Maybe the back room boys will figure it out when we get home. If not, the Beehive asteroid will not pass through the inner Solar System for another one hundred and thirty-seven years."

"I'd like to think we will have ships capable of visiting it long before then." Diane turned to Pierre. "But at least we've answered one question."

"Do you really believe what you told Poornima?" Marvin asked.

In response, Pierre asked the computer to project an image over the table. The hexagonal cavity with its six stones hovered holographically before them.

"The stones were all perfect spheres. When I got them back here, I measured and massed them." He pointed. "Starting from that one, the smallest, and going clockwise, each subsequent stone weighs exactly twice as much and is double the diameter of the one before."

"A binary sequence," Diane whispered.

Pierre nodded.

"That's it?" Marvin asked. "Just...stones?"

"Oh, Marvin, of course not 'just stones'," Diane admonished gently. "It is information. An...an artefact of the *mind.*"

"What are they trying to tell us?" Marvin wondered.

Pierre shrugged. "There is no specific message per se. As Diane said, it is information, like that early SETI radio message that was beamed to the Hercules Cluster—pictures of DNA and things like that. It just shows an understanding of some fundamental concepts. It shows...intelligence."

"Intelligence," Marvin repeated.

"But there is so much we don't know," Diane said with frustration. "We cut and run just as we start making discoveries. We find some answers, but end up with more questions instead." She clenched her fists. "It's the nature of government science!"

"No," Pierre said quietly. "It's the nature of exploration."

The intercom came on. "*Three minutes to TEI burn. I'm sure you don't want to miss this.*"

"Thanks, Ben," Pierre said.

"*It's my job,*" Ben replied. "*Oh, one more thing. Darmstadt tells us the IAU has officially renamed 2046-PK as asteroid 329780 Beehive.*"

"Departure. So soon." Diane started for the door. "Aren't you two coming?"

"I am." Marvin got up, but Diane had already left the room. He went as far as the door, and stopped.

"Pierre?"

"Yes."

"About those stones." Marvin paused for a moment in thought. "You know…when people encounter something they've never seen before, something totally unknown, there's sometimes a tendency—maybe subconscious, but it's there—a tendency…to try to relate it to ourselves, to something familiar, something we know about, because we're so desperate to understand it."

"And?"

"I guess what I'm saying is…Are you sure you're not just seeing what you want to see?" He started for the door.

"Marvin."

He paused.

"I don't think so."

Marvin shrugged, and left the room.

Pierre rested his elbows on the table, his hands cupped under his chin. The ghostly image of the six stones hovered before his eyes: the stones, the binary sequence, the message. He still had no idea who or what the aliens were, but through a common informational choice, a link had been forged between himself and the builders of the Beehive. They proved that all sentient beings, regardless of biology, shared at least one common heritage. Pierre marvelled once more at this simple message, transmitted across an unimaginably vast expanse of space and time, from a stone.

Afterword

Much of the science in "From a Stone" came from a planetary geology course taught by the late Professor Linas Kilius that I took during my undergraduate studies at the University of Toronto. This course introduced me to fundamental concepts such as the different classifications of asteroids and the technique of crater counts to estimate the relative ages of planetary surfaces. I was also greatly influenced by *A Man on the Moon*, Andrew Chaikin's superb history of the Apollo program.[1]

The fictional asteroid in "From a Stone" was a C-type with a spectral classification similar to carbonaceous chondrite meteorites. C-type asteroids are the most common in the Solar System, forming about 75% of the known population. They are amongst the darkest objects in the Solar System, distinguished by a very low albedo (propensity to reflect solar radiation) due to their high concentration of carbon. This makes them difficult to detect, so in the story it was fortunate the United Nations Space Development Agency had the Shapley Observatory on the airless lunar farside. C-type asteroids are most commonly found at the outer edge of the asteroid belt between Mars and Jupiter, about 3.5 astronomical units (525-million kilometres) from the Sun. The notion that the asteroid in the story was in a highly elliptical orbit that took it beyond the Kuiper-Edgeworth belt (30–50 astronomical units) made it a most unusual C-type, perhaps hinting at the alien artefacts waiting inside for the crew of the *Harrison Schmitt* to find.

Counting craters to estimate the relative ages of planetary surfaces was pioneered in the 1960s by a number of scientists including Ernst Öik, Gene Shoemaker, Robert Baldwin, and Bill Hartmann. When the surface is new, it is assumed to have few or no impact craters. Over time, the surface is expected to accumulate impacts based on rates of bombardment that are assumed to be known for various epochs of the Solar System. The more heavily cratered the surface, the older it is likely to be. This technique is obviously not useful for planets like the Earth with active surfaces that are reshaped by plate tectonics, volcanism, and weathering. Radiometric dating of rock samples returned from the Moon has been used to calibrate the crater counting method, and in the near future the same could be done with samples from asteroids and Mars. The size, distribution, and state of degradation of craters also provide clues to the geological history of a surface. Pierre Caillou's curiosity was piqued when he saw a newer crater appearing more degraded than a larger and older one underneath it.

The "artefact of the mind" of the six stones was drawn from an article by Phylis Morrison in the January/February 1992 issue of *The Planetary Report* in which she described an exhibit at the Boston Museum of Fine Arts consisting of seven rounded river pebbles arranged by increasing size. These pebbles originated in the Indus Valley civilisation of what is now Pakistan and were over 4,000 years old. Each pebble weighed twice as much as the one before, resulting in a binary sequence capable of representing every integer

---

[1] Chaikin, Andrew. *A Man on the Moon: The Voyages of the Apollo Astronauts*, Penguin Books, 1994.

from 1 to 128 (i.e. 2^7) to be used as units of measure in the pan balances of Harappan merchants. Much as Pierre Caillou contemplated the mathematical connection between himself and the aliens of the Beehive asteroid, so too did Phylis Morrison conclude her article by marvelling at "the informational choices…[that] linked me to those clever artisans of the past."[2]

Most people are familiar with SETI (search for extra-terrestrial intelligence) from the standpoint of monitoring the electromagnetic spectrum for radio or optical transmissions from civilizations outside the Solar System, as popularized by the Carl Sagan novel (and later film) *Contact*. Some have suggested that SETA, a search for extra-terrestrial artefacts, might be a useful complement to radio and optical searches.[3] The argument has been made that conveying a message across interstellar distances with a physical artefact like a spacecraft (or a tunnelled C-type asteroid) might be more energy efficient than electromagnetic communications if travel time is not a consideration. Similar to the concept of the "water hole" frequencies of the hydrogen and hydroxyl lines in radio SETI, it has been suggested that gravitational "traps" such as the Sun-Earth or Earth-Moon Lagrange points could be telescopically searched for potential artefacts. Humans have included messages on five spacecraft that have left the Solar System, consisting of plaques aboard Pioneers 10 and 11 as well as the famous golden records on Voyagers 1 and 2. A digital message will be uplinked to the New Horizons spacecraft after the completion of its mission through the Kuiper-Edgeworth belt.

"From a Stone" was first published in the September 1996 issue of *Science Fiction Age* magazine. The editor Scott Edelman suggested revising my first draft to have the spacecraft and its crew at the end their mission and therefore constrained by resource limitations (and orbital dynamics) to having only a brief time to study the asteroid. Not only did his suggestion increase the dramatic tension, it also created a more realistic portrayal of the compromises inherent in trying to do science and exploration within the practical and political constraints of a government space program. Only one of the twelve people who walked on the Moon during the Apollo program was a scientist—the geologist Harrison Schmitt, for whom the ship in the story is named. Proponents of astronauts have argued there is no substitute for the adaptability of a human scientist to unexpected situations in the field, yet at least in the Apollo experience with "flight plans as tightly scripted as Shakespeare" (a great line from Canadian journalist Allen Abel) there was

---

[2] Morrison, Phylis. "Pebbles on the Massachusetts Shore", *The Planetary Report*, Jan/Feb 1992, Pg. 18.

[3] Freitas, Robert A. "The Search for Extraterrestrial Artifacts (SETA), *Acta Astronautica*, 12(1985):1027–1034, IAA-84–243(A).

actually little time for either spontaneous discovery or detailed study. The advantages and disadvantages of human versus robotic space exploration is a topic I explore further in the title story "Just Like Being There".

The Coming Age of the Jet

For my father Douglas Choi, another hero of Canadian aviation.

*There is no telling what Avro would have achieved in the commercial field if the
company had been permitted to go ahead. The engineering genius which produced the
Jetliner went on to create the CF-100 and the spectacular Arrow. If that genius had
been applied to the Jetliner's successors, the world's airlines might long before now have
been beating a path to Canada's door, instead of placing their orders with
manufacturers in the United States and Britain [where even Canadian airlines must
buy their equipment].*
—The Globe and Mail, 26 September 1958

Malton, Ontario, Canada
September 1950

The Experimental Section of A. V. Roe Canada's Malton plant, the birth-
place of the C102 Jetliner, resembled nothing less than a graveyard. The area
was almost empty, with only a few pieces of abandoned equipment left about.
On most days, the sorry state would have saddened Stanley Dixon. But this
morning, he felt differently. Since Avro management considered it an eyesore,
visitors were seldom shown the area now. Today, that suited Dixon just fine.

Peering out from behind a crate, he made sure the corridor was empty
before approaching the office of Joyce Rigby, one of the Jetliner project's
engineering draftswomen. He knocked.

"Come in."

"Hi, Joyce. I was wondering if…"

© The Author, under license to Springer Nature Switzerland AG 1997
E. Choi, *Just Like Being There*, Science and Fiction,
https://doi.org/10.1007/978-3-030-91605-3_5

Rigby was cleaning out her desk.

"Oh, no. Not you too."

Rigby nodded. "I'm afraid so, Stan. I've been –"

"– transferred to the CF-100 project."

Both spoke the words together. To Dixon, this was emotionally the hardest part. The Jetliner team, his friends and colleagues, were being systematically drained off to the military fighter programmes at an alarming rate. He never knew who might be gone the next morning.

"Well, good luck."

"Thanks." She closed her drafting board. "Did you want something?"

"Oh, yes. I was wondering if you've finished those drawings I asked for."

"Of course. They're right here." Rigby went to a filing cabinet and handed Dixon a folder. "Still producing proposals for the airlines?"

"I'm not giving up."

"I know," she said quietly.

Dixon leafed through the schematics. "They're perfect. Thanks Joyce."

"You're very welcome. Any time, Stan."

As Dixon turned to leave, Rigby spoke again. "By the way, did you know Minister Howe's touring the plant this morning?"

"Uh, huh. Checking up on his precious fighters again." Dixon lowered his voice. "I've been trying to *avoid* him."

Rigby smiled. "You're much too young to be so cynical. But good luck anyway."

"Thanks."

Slipping back through the bone yard of Experimental, Dixon thought he'd returned to his office without detection. But just as he reached his door, a voice suddenly called out from behind.

"Stan!"

Dixon whirled.

Fred Smye, the general manager of Avro's Aircraft Division, was waving him over. At his side was Clarence Decatur Howe, the right-honourable Minister of Defence Production.

Smye smiled. "Stan, there you are. I've been looking for you all morning. I wasn't going to let Mr. Howe leave without meeting you."

"That's very thoughtful of you, Fred." Dixon turned to the minister. This was the first time Dixon had met him in person, but he looked much as he did in the newspaper photos. C. D. Howe was a short, rather obese man in a conservative grey suit and a homburg hat. His hair was completely white, and his wrinkled face wore a perpetually sour expression.

"Mr. Howe, this is Stanley Dixon. He was one of the engineers who designed the Jetliner. He's currently our chief sales representative for that aircraft."

"Jetliner, eh?" Howe scowled. "Well, I suggest that you forget about that airplane and put your energy into getting the CF-100s out!" He abruptly turned and walked away. Smye looked at Dixon, shrugged his shoulders, and followed the departing minister.

Dixon's jaw almost fell to the floor. Under the leadership of chief engineer Jim Floyd, dedicated men and women like himself and Joyce Rigby had given their hearts and souls for the Jetliner. Through their efforts, Avro had produced a world-beating aircraft—an aircraft that would now likely never show its true potential to the world. For four years of blood, sweat and tears, all this politician had to say was "forget it"?

He was in a bad mood for the rest of the day.

* * *

Sitting on the porch of his modest home, Dixon sipped a glass of lemonade, watching some kids play hockey in the street. The sight of children made him forget a little the upsets of the day, and increased his anticipation of becoming a father himself.

Hearing the screen door open, he turned to see his wife Rhoda. Dixon put down the lemonade and embraced her.

"How's our prototype?" He rubbed her belly.

"Fine." She frowned. "You didn't have a very good day, did you?"

He shook his head.

"Want to talk about it?"

Dixon's closed his eyes. "Do you remember when the war ended, and I got that form letter from Victory Aircraft? The one that thanked me for a job well done—and then asked me to seek other employment?"

Rhoda nodded. "But Larry Marchant asked you to stick around after Avro took over...and here you are, still sticking around."

"The Jetliner's being put on the back burner. I think they're going to kill it."

She pulled her husband closer. "But what's the worry? Avro's got all those fighter planes to build. You won't lose your job."

"No, it's not just that. This is *wrong*, Rhoda. What the government really wants is for us to abandon the Jetliner to concentrate on their military projects. I think Dobbie and Fred know, but they're not doing anything." Dixon shook his head. "Relying solely on government contracts is a *bad* idea. We'll live to regret it."

Rhoda ran her fingers through his hair. "Stan, I'm sure everything's going to be fine. You've had a bad day, but it's over now. Just forget about it."

Dixon grimaced, but quickly stifled it with a smile.

"Dinner will be ready soon."

The Jetliner's problem—Dixon's problem—was Trans-Canada Airlines. Avro had designed the plane to the specifications of the government-owned carrier, but TCA president Gordon McGregor later decided that his airline should not take the risk of being the first to operate jets. Not only did they refuse to buy the Jetliner, they also went on to make a series of negative public statements about the plane, apparently in an effort to justify their decision. That hurt Dixon's marketing campaign, since the first question potential customers asked was why the national carrier had rejected an aircraft they had originally specified.

The tranquillity of the late afternoon was interrupted by the drone of propellers. Dixon looked up just in time to see a Convair 240, its wheels down for landing, dart over the house. The plane sported a stylized Canada goose logo on its tail, with blue, white, and red stripes along the fuselage. Dixon sighed, wondering how things might have turned out differently if Avro had approached *that* airline instead of TCA.

An idea came to him. "Military projects…"

Perhaps it wasn't too late.

Dixon ran into the house. Startled by the slamming of the screen door, Rhoda emerged from the kitchen to investigate.

"Dinner's…Stan, what are you doing?"

"I have to pack."

"Why?"

"I'm going on a trip."

"A trip? Where?"

"Vancouver."

* * *

Grant McConachie, the president of Canadian Pacific Airlines, sat behind an ornate ebony desk, laconically smoking a cigar. The desk was littered with models of various airplanes in CPA livery. Framed photographs of other CPA equipment and mementos from his bush pilot days adorned the walls. Behind him, a set of clocks displayed the time in four zones.

McConachie rose and shook Dixon's hand firmly. His round open face, which still bore the freckles of youth, was crowned by a shock of sandy brown hair with just the slightest hint of grey.

"So, you're the sales person from Avro."

"Yes."

"Well," McConachie said, "I was rather hoping that one of your engineers would come out and –"

Dixon interrupted. "Actually, I'm an engineer."

"Really?" McConachie broke into a wide grin. "Super! There's something I want to show you."

Before Dixon could say another word, McConachie suddenly whipped out an inflatable plastic globe and began blowing it up.

"What...uh, what are you doing?"

"Just a moment...just...a bit here..." McConachie was trying to talk in between blowing up the globe and smoking the cigar. "Only takes a minute..."

After inflating the globe, McConachie opened a drawer and produced a piece of string. "All right, now. Tell me Mr. Dixon, what's the shortest distance between two points on a sphere?"

Dixon shrugged. "An arc."

"Yes! A Great Circle!" He pulled the string across the globe, connecting Vancouver and Amsterdam across the North Pole. "Do you know what this means?" Dixon opened his mouth, but McConachie continued. "From the west coast of Canada to Europe, the Great Circle is a thousand miles shorter than conventional routes via Montréal or New York. Passengers will pay the same fare but fly less, thus boosting our revenue yield per mile flown." He sighed. "Now, if only the damned government would give us the approval."

With that, McConachie released the valve—and then sat on the globe to deflate it. As the hot air and cigar smoke evacuated, it made a noise that sounded rather obscene. He smiled sheepishly. "Anyway, what can I do for you?"

"Mr. McConachie, I want to sell you an airplane. A *jet* airplane."

* * *

Although the two men talked the remainder of the afternoon, little of it was about the Jetliner. Instead, Dixon found himself somewhat overwhelmed by McConachie's seemingly endless stream of anecdotes: Shovelling coal as a teenager to pay for flying lessons; rescuing two brothers in northern Alberta burned by an exploding stove; making the first commercial flight over the Rocky Mountains from Calgary to Vancouver; becoming CPA president at the age of 37; going to Tokyo to personally ask General MacArthur for permission to fly to Japan; battling Ottawa for the right to full and equal competition with TCA. The day faded into evening, and at McConachie's suggestion the two men went for dinner at the Chartwell Restaurant. As the meal progressed, conversation returned to the matter of the Jetliner.

McConachie took a sip of wine. "You know, of course, that I've already bought a jet airplane."

"CPA has two de Havilland Comets on order. I'm aware of that."

"So what can you offer me that de Havilland can't?"

"A chance to be the first airline in North America—and one of the first in the world—to operate jets."

"Go on."

"Those Comets won't be delivered for another two years. I can give you a Jetliner later this year."

"How can you give me a Jetliner if you don't even have a production line yet?"

Dixon took a deep breath. "We'll refurbish the prototype and loan it to CPA."

"You'll lease me the prototype?"

"No, sir," Dixon said slowly. "The prototype will be loaned for CPA's use— at no expense—until the production models are ready." The passive voice was deliberate. Like this meeting with McConachie, he had not cleared that little incentive with anyone at Avro.

"I see." McConachie took a bite of salad. "You've got a super airplane. I'm very impressed. But I'm afraid it's not quite appropriate for our routes. If only your company had come to us sooner, we could have…Well, in any event, what we really need is a long-range jet like the Comet."

"Point taken. But if we use the Allison J33 engine and put auxiliary fuel tanks in the fuselage and outer wings, the Jetliner can match the Comet's range. As is, the Jetliner is ideal for your Canadian services. The Comet's too big to fly domestic economically."

McConachie pursed his lips. "You know, what I'd really love to do is fly a jet from Honolulu to Sydney. Of course, it'd have to refuel in Fiji and Canton Island, but the speed should more than make up for the stops. Certainly, I need something to replace those damned Canadair Fours!"

The Canadair Four was a copy of the American DC-4, commissioned by the Canadian government in an attempt to foster the domestic aircraft industry. It was heavy, had a voracious appetite for fuel, and produced a horrendous noise that no amount of soundproofing could dampen. On top of that, the short inboard exhaust stacks of its Rolls-Royce Merlin engines occasionally bellowed blue flames, terrifying the passengers.

Dixon saw his opening. "Actually, I was thinking of Vancouver to Tokyo."

"Vancouver to Tokyo? How?"

"Vancouver to Anchorage, Anchorage to Shemya, Shemya to Hakodate, and then on to Tokyo."

"Interesting." McConachie rubbed his chin. "And why, Mr. Dixon, should CPA fly your Jetliner on the Vancouver-Tokyo route?"

Dixon cut a piece of sirloin and held it up with his fork. "You convinced Ottawa to give CPA a contract to airlift American military personnel to Tokyo for service in Korea. That contract requires you to provide only the meagre essentials. Anyone on a U.S. transport would get a canvas bench and a cold sandwich. But you serve steak and champagne. Why?"

McConachie shrugged. "Word of mouth advertising. Today's military officer is tomorrow's executive. We want those people talking us up."

Dixon smiled wickedly. "Well, imagine how much talking they'll be doing after having steak and wine at *four hundred miles an hour*."

McConachie lit a cigar. "I can imagine a lot."

* * *

Dixon had barely put down his briefcase when Mario Pesando, the Jetliner's chief of flight testing, poked his head in the door. "*You* are in *big* trouble!" He jerked his thumb. "Dobbie wants to see you in his office, right away."

With large puffy cheeks and sad drooping eyes, Sir Roy Hardy Dobson, the president of A. V. Roe Canada, looked very much like a bulldog. The soulful eyes tracked Dixon as he entered the office and took a seat.

"I understand," said Sir Roy, "that you took a trip to Vancouver last week."

"I did," Dixon replied. "But it was on my own time and my money."

"Perhaps. But you were acting as if you were a company representative, conducting business you had not consulted me on."

Dixon said nothing.

"Stan, you know as well as I that we cannot afford to waste any more time on the Jetliner. The CF-100 is our first priority."

Dixon leaned forward in his seat. "We have a good airplane, Dobbie. Years ahead of anything else. All we need is *one* customer, one person with the guts to be first. After that –"

"We *have* a customer," Sir Roy interrupted. "The government needs those fighters, and we must concentrate our efforts to that end."

Dixon shook his head. "We can't rely on government work, Dobbie. That's an invitation to financial suicide."

"Nonsense. With the CF-100 and the supersonic interceptor project, this company's future is assured."

Dixon said, "Then what should I tell CPA?"

"I beg your pardon?"

"Grant McConachie is coming to Malton to see the Jetliner tomorrow. I told him we'd take him on a demonstration flight."

"Stanley, what have you done!" Sir Roy exclaimed. "Minister Howe stated quite explicitly that there should be no more Jetliner demonstrations. We shan't defy that edict."

"Why not?"

"Because the government is buying seven hundred and twenty CF-100s with one hundred percent spares. They expect us to produce twenty-five fighters each month. That's a lot of leverage, Stan."

Dixon took a deep breath. "This isn't about the Jetliner, this isn't about engineering, this isn't about business. It's about Chadwick, isn't it?"

The Avro president closed his eyes, and for several moments did not speak. Roy Chadwick, his lifelong friend and associate, was killed in a crash of a propeller-driven Tudor airliner two years ago. From the start, the Tudor seemed a cursed project, with the British government running interference and the state airline BOAC refusing the buy it. Following another Tudor crash last year, the British Minister of Civil Aviation banned the type from passenger service.

"He was my friend too, Dobbie. And I know that you've always associated civil projects—the Jetliner especially—with him." Sir Roy opened his mouth, but Dixon cut in. "Please let me finish. Chadwick was my friend, and he was a damned fine engineer. But do you think he'd want this? Do you think he would want us to throw out four years of blood, sweat, and tears? Do you think he'd be happy that we were turning our back on this great achievement, this great airplane?"

After several long moments, Sir Roy spoke. "This McConachie fellow, I hear he's quite a character."

"You could say that, yes."

"I think I'd rather like a man like that." The Avro president forced a weak smile. "I suppose we shan't disappoint him."

Dixon smiled back. He would tell Sir Roy about the prototype deal another time.

* * *

The Avro C102 Jetliner, registration CF-EJD-X, sat on the apron in front of the Avro plant, basking in the early afternoon Sun. With even a cursory inspection, it was obviously different from any kind of airplane that had come before it. Its fuselage was sleeker and more streamlined, from the rounded point of its nose to the graceful arch of its tail. But the most telling difference was the absence of propellers. Instead, the plane was powered by four Rolls-Royce Derwent turbojets clustered in pairs under the base of each wing.

Grant McConachie walked in endless circles about the Jetliner, methodically scrutinizing the plane. "…low slung engines, easy to maintain…just slide them out…"

Dixon cleared his throat. "Uh, Mr. McConachie? We're ready to fly."

McConachie looked up. "Oh, yes. Of course."

Once aboard, Dixon introduced McConachie to the crew. For this flight Don Rogers was the captain, with Sid Howland acting as the co-pilot. Dixon himself would monitor the flight engineer's console, while McConachie took the spare jump seat.

"*Avro EJD, you are cleared for take-off.*"

The engines revved up from a low hum to a soft roar, and the Jetliner streaked down the runway. Reaching take-off speed, Rogers pulled back on the control column, and the Jetliner soared into the sky. Upon reaching altitude, the engines were throttled back to their cruise setting.

McConachie looked about. "It's so quiet. You guys don't even use earphones up here." He unbuckled his seat-belt and darted into the passenger cabin, returning a few moments later. "It's quiet back there, too. Did you put any soundproofing in?"

Dixon shook his head. "Some, but it's not complete."

"There's almost no noise, no vibration!" McConachie raved. "What's our airspeed?"

"351 miles-per-hour," replied Sid Howland.

"This is super!"

"Mr. McConachie," Captain Rogers said, "Would you care to fly her?"

"Oh, yes. Thank you."

Captain Rogers let McConachie have his seat. The former bush pilot deftly took the controls. For the next two hours, the CPA president put the Jetliner through its paces. Throughout it all, he wore a huge grin on his face.

At last, Rogers tapped him on the shoulder and pointed at the fuel gauge. "Time to head home, sir."

Reluctantly, McConachie relinquished the captain's chair. Turning the control column, Rogers put the Jetliner into a wide banking descent that brought them back towards Malton. As they lined up with the runway and entered final approach, co-pilot Howland pulled the lever to deploy the landing gear.

The three lights on the panel remained red.

"Looks like we're having a little trouble with the undercarriage," Rogers said.

Howland tried the lever again. Nothing.

"*Avro EJD, do you wish to declare an emergency?*"

"Standby on that," Rogers told the controller, "until we've characterized the malfunction."

The Jetliner aborted the landing and circled the airport. As it passed by the tower, the controller came back on. "*Avro EJD, we see your wheels are up.*"

"Copy, Malton," Rogers said. "Well, it's not a problem with the lights then."

"The electromechanical selector doesn't seem to be working," Howland reported. "I'm going for manual deployment." He knelt down and removed a panel on the floor, exposing a small lever. Its purpose was to release hydraulic pressure, opening the undercarriage locks. If these could be disengaged, the wheels should drop into the slipstream and lock down.

Howland pulled on the lever without effect.

"Try again," Rogers suggested.

The co-pilot heaved with all his strength. Suddenly, something snapped. Dixon heard a thud.

"Nose gear down," Rogers reported. "But the mains are still up."

Howland returned to his seat. "Unfortunately, that cost us hydraulic pressure. Flaps and ailerons are inoperable."

"This is Avro EJD, we are now declaring an emergency," Rogers said.

"*Affirmative, Avro EJD. Traffic has been diverted, the airport is yours. Good luck.*"

Rogers lined up the Jetliner with the runway. After crossing the threshold lights, he carefully throttled back the engines, trying to reduce speed without stalling. As the nose wheel touched the ground, Rogers struggled to hold the rest of the plane up until the last possible moment. It finally set down with an ear-piercing screech of metal as the engine nozzles dragged along the runway. Moments later, the plane came to a halt.

Howland and Rogers were sweating. McConachie was rubbing a bump on his forehead. Dixon hung his head in despair. Everyone was safe, but the Jetliner's fate was sealed.

* * *

Stanley Dixon found himself in Sir Roy's office the following afternoon. But this time, they were not alone.

C. D. Howe was slouched in an armchair, his legs man-spread. The minister glared at Dixon as he entered the room. After a few moments of awkward silence, Sir Roy spoke. "Stan, what is the condition of the Jetliner?"

"Some slight damage to the underside of the nacelles and rear fuselage. Very minor. It won't take long to fix."

"Do you know what happened?"

"One of the liquid-spring shock absorbers failed, jamming the undercarriage. Design modifications are being made. It won't happen again."

Howe cleared his throat. "Well, I should certainly hope a lesson's been learnt from this. The C102 is a failure."

Dixon said, "Now wait just a —"

Howe turned to Sir Roy. "Dobson, I *told* you that your only priority is to get those fighters out. That golden fleece of a Jetliner is simply in the way."

"That's not true!" Dixon snapped.

"Isn't it? Have you ever heard of someone who has too much on their plate? Well, Avro has too much on its plate in trying to do both the fighter program and the Jetliner program. This government is committed to those fighters. The Jetliner must go on the shelf." Howe paused to catch his breath. "In any event, it cannot be denied that the Jetliner is a technical failure. Everything is wrong with it. I understand you even have to carry sand in the tail to make it fly, like you did yesterday."

With nominal patience, Dixon explained, "We put sandbags in the *cabin* to simulate the weight of the passengers."

"Yes, yes." Howe waved dismissively. "But my independent consultants assure me the Jetliner is wholly unsuitable for commercial service."

Dixon groaned. Howe's "independent consultants" were officials from Trans-Canada Airlines. "I'll never understand," he said, "how people like you can so easily crucify the truth for the sake of political gain."

"Stanley!" Sir Roy exclaimed. "You will not address the honourable minister in that manner!"

"You will not address me in that manner!" Seething with anger, Howe continued. "I had heard rumours that Avro was planning to use part of the space in this plant for further work on the Jetliner and other commercial ventures. Your foolish demonstration flight yesterday realized my fears. Having in mind the colossal investment of government funds into this company, I will not tolerate *any* such use of your floor space. As of this moment, I am formally *ordering* you to terminate all civilian work. Since the Jetliner will be supporting the CF-100 flight tests, the repairs may proceed. But when those flight tests are over, that cursed airplane shall be *scrapped*."

Heaving himself out of the chair, Howe waddled over to the rack to pick up his homburg. "Be assured that you will get official written notification in the very near future. Good day, gentlemen."

He slammed the door behind him.

* * *

Dixon knocked on Room 629 of the Royal York Hotel in Toronto.

"Mr. Dixon," Grant McConachie said upon recognizing his visitor. "Come in."

"I just wanted to come by and see you before you left."

"Thanks."

Dixon pointed. "How's your head?"

McConachie rubbed the spot. "Better."

"I see." Dixon began to fidget. "You're not going...you're not going to, you know..."

"Well of course I am!"

"Oh." Dixon's heart fell. "Well, I suppose we'll have to call our lawyers."

"Unfortunately, yes. I'm afraid our lawyers will have to get involved."

"Mr. McConachie, I'm sure we could settle this out of –"

"I want three of those Jetliners."

Dixon's jaw fell. "What did you say?"

McConachie picked up his inflated globe. "I've spoken with our board of directors. It was a tough one, but I sold those bastards." He began tossing the globe in the air. "Mr. Dixon, I want three Jetliners to complement our Comet fleet."

Dixon was at a loss for words. "But...but after that accident, I thought –"

"Oh, that." McConachie shrugged. "You've got a super airplane, Mr. Dixon. I want it. Forget about that scrape up. I've been through so many of those in my bush days, I've lost count. Actually, that little incident showed me another advantage of your jet. If that plane had propellers, the damage would have been severe. Besides, think of it as one more certification test you won't have to do."

Dixon's momentary joy was quickly extinguished by cold reality. "Mr. McConachie, I'm afraid that won't be possible."

"Why not?"

"The government has decreed that Avro abandon its commercial ventures and concentrate solely on their military projects. I'm sorry, but there will be no production of the Jetliner."

"I see." Still playing with the globe, he continued. "Well, I've come to expect things like this. Believe me, I know how you feel. I've got my own problems with the government. Do you know how they assign routes?"

"CPA is only offered routes already rejected by TCA as unprofitable."

"That's right. Do you know what I do?"

"What?"

With a sudden snap of the wrist, McConachie tossed the globe across the room. Dixon was startled, but managed to catch it.

"I play the hand I've been dealt, no matter how bad the cards seem. Always make the bold move first, my friend, and things will find their place." McConachie lit a cigar and flashed a confident grin. "I know why you were so keen on having us fly the Jetliner to Tokyo. It's a good idea. It just might give us the leverage we need. We'll see."

Stan Dixon just stood there, holding the world in his hands.

* * *

"*One minute to launch.*"

Dixon peered out the Jetliner's window to look at the other plane flying alongside. Despite having no real business being here, he'd managed to talk his way aboard. He seized every chance to fly the Jetliner now, knowing that each flight might be the last.

That other plane was a CF-100 Canuck fighter. It was an ungainly creature, with blunt nose, straight wings, and two Orenda M11 engines stuck to the sides of its cigar-shaped fuselage like an afterthought. Mounted under the belly was a rocket pack, the subject of today's weapons compatibility test.

Dixon sighed. The Jetliner—and its successors—should be lofting men, women, and children to the four corners of the globe. Instead, it was relegated to the role of a photographic platform for the CF-100. Ironically, it was the Jetliner's speed and altitude capabilities that made this reprieve possible. But afterwards, Dixon knew, this beautiful airplane will end up as pots and pans.

"*...three...two...one...fire!*"

With a flash of flame, a salvo of missiles was unleashed from the CF-100, trailing thin columns of smoke. Dixon heard cameras clicking accompanied by the cheering of the officers from the Royal Canadian Air Force.

"Okay, it looks like we've had a good test," said Captain Don Rogers over the intercom. "Heading back to base."

The Jetliner banked and pulled away from the CF-100. Fifteen minutes later, it touched down on the runway at Camp Borden. The landing gear worked perfectly.

As Dixon walked across the tarmac, he was intercepted by a young lieutenant.

"Mr. Dixon?"

"Yes."

"Please come with me, sir. There's a telephone call for you."

Dixon followed the lieutenant to a briefing room, where he was shown a telephone on the wall. He thanked the young officer and picked up the handset.

"*Stan, how are you?*" It was Sir Roy.

"I'm fine, Dobbie."

"*How was the flight?*"

"It went well."

"Good. Stan, I've some important news for you. You'll get a more formal notification soon, but I wanted to tell you in person. As of next week, you will no longer be the chief sales representative for the Jetliner."

It took a moment for that to sink in. "I see." He thought about Rhonda, the baby—and a form letter. "Dobbie, I know we've had our differences, but...I would be most grateful if you might find me another position in the company."

"I've already taken care of that. You will be reassigned as manager of production engineering."

Dixon exhaled slowly. "Thank you, sir. I'm looking forward to working on the CF-100."

"The CF-100? Why the devil would I want you working on that? I want you to be manager of Jetliner production."

Dixon gulped and blinked furiously a few times. "The Jetliner, sir?"

"That's correct. CPA's signed a contract for three firm orders, plus two options. The government's approved it, on the condition that we finance the plant expansion ourselves. Avro is building the Jetliner, Stan. I want you in charge."

Four cadets having lunch in the cafeteria looked up from their meals, wondering who the hell was making so much noise down the hall.

* * *

Vancouver, British Columbia, Canada
January 1951

The afternoon before the roll-out ceremony at Vancouver Airport, Dixon paid Grant McConachie a final visit at CPA Headquarters.

"Mr. Dixon, how curious that we're always bumping into each other like this," he said, toying with a model of sleek, delta-shaped craft with rocket pods under the wings. It was, of course, in Canadian Pacific colours.

"I have something for you." Dixon reached into his pocket, fished out a small metallic object, and handed it to McConachie. It was the symbolic key to start the plane.

"You once told me something about playing the hand you're dealt. Well, sometimes I still can't believe the pot Avro's collected." Dixon ticked his fingers. "The Jetliner's gone from being a government orphan to...oh, how did they word it? 'An essential component of Canada's U.N. commitment to Korea.' We've got orders from National and Eastern, we're negotiating with American and United...even TCA wants to talk again."

"And we were right about Tokyo."

"The U.S. Air Force and Navy are both interested in the Jetliner as a troop transport. Your Korean airlift gives them a no-risk, no-cost way of evaluating the plane in that mission. So, we got their support too." Dixon paused. "Thank you, Mr. McConachie."

"Thank *me*? Whatever for?"

"For having the guts to be first," Dixon said. "But also for...well, I've heard certain rumours."

"About what?"

"Oh, that C.D. Howe got a call from General MacArthur himself. Something about the absolute necessity of getting his men to Korea by the fastest means possible." Dixon looked McConachie in the eye. "You wouldn't know anything about that, would you?"

McConachie lit a cigar. "Nope."

Dixon smiled and pointed at the model. "What's that?"

"This?" McConachie picked it up and swiped it through the air. "Imagine taking off from Montréal at dawn, flying west, out speeding the Sun. You could look back over your shoulder to see the newly-risen Sun setting in the east as you cruise westward into the night. Or...I'll bet you could fly from San Francisco to Tokyo in twenty-seven minutes. Right up to outer space, and then back down again. Hmm...that's a pretty wild ride. Maybe the passengers will have to be put in a horizontal position. I'll have to think about that."

McConachie held up the rocket plane. "Do you think your company could build me one of these?"

* * *

The sky above Vancouver Airport was overcast, with low clouds hanging overhead like giant wads of steel wool. According to the meteorologists, there was a forty percent chance of rain. It wasn't a good day for a ceremony, but Dixon knew they'd press on anyway. A *Globe and Mail* newspaper reporter once called McConachie an "irresistible force". The weather would hold.

"This is an epic-making event in the story of Canadian aviation. The C102 Jetliner is an aircraft conceived, designed, and developed in Canada by Canadian engineers and fabricated by Canadian workmen. It is the first all-Canadian airplane to be powered with jet engines. More than that, it is the first transport plane using jet engines ever flown on the continent of North America. Surely, this is a great achievement for Avro, and for every man associated with that company. I am sure that I speak for all those present when I wish every success to the C102, its manufacturer A. V. Roe Canada, and its first customer, Canadian Pacific Airlines."

Dixon joined the polite applause as C. D. Howe left the podium. He was sitting on a raised platform in front of the audience along with other company dignitaries like Jetliner designer Jim Floyd, Fred Smye, Sir Roy Dobson, and the new president of Avro Canada, Crawford Gordon. Further down were the CPA officials, including chief engineer Ian Gray, treasurer Harry Porteous, and of course Grant McConachie himself.

Dixon estimated there were several hundred people present. They were mostly CPA employees and reporters, but a few Avro people were present. A waving arm caught his attention. Turning in that direction, he spotted Rhoda with Joyce Rigby. He smiled and waved back.

A hush fell over the crowd as McConachie took the podium.

"I want to thank you all for coming out on this rather dreary day. I am delighted to be here, and I'm particularly delighted to be presiding over this very special occasion.

"Some four hundred years ago, Magellan took a thousand days to sail around the world. Three centuries later, after the invention of steam power, history was made with a voyage around the world in eighty days. Today, after just half a century of aviation, the globe can be circled in eighty hours.

"If Magellan's world could be equated to a watermelon, it took three hundred years and steam power to shrink it to the size of an apple. But in half a century the airplane has shrivelled it to the size of a pea, and in another fifty years will reduce it further to the size of a pinhead.

"Unfortunately, we're not at the pinhead stage yet...although some have said that I'm already there!" A chuckle went through the crowd. "But we are clearly on our way, and today will be recorded in history as the start of a new era in transportation. Canadian Pacific has always had a proud tradition of leadership. This airline and this nation must not—will not—founder in the backwash of the coming age of the jet. Thanks to the engineers of Avro, we have been given an opportunity to be pioneers of this new age. CPA intends to seize this opportunity. We have the best people, the best service, and the best technology to serve Canadians from coast to coast and around the world, providing a true, private enterprise alternative to...that other airline."

The audience laughed and cheered. McConachie broke into a wide grin. "Ladies and gentlemen, I present to you…Canadian Pacific's brand new, Super Jetliner!"

The hangar doors parted. With deliberate majesty, the Jetliner was slowly towed out. When the tail cleared the building, the tug made a turn to bring the side of the fuselage within view of the crowd. The refurbished prototype still bore the registration CF-EJD, but the experimental "X" suffix was no more. Gone also were Avro's white and yellow in-house colours. It now wore the blue, white, and red stripes of CPA. "Canadian Pacific" was written in script above the row of windows, and the tail bore the stylized CPA Canada goose insignia. At the front of the plane, in simple but proud block letters, was Jetliner CF-EJD's new name: *Empress of Canada.*

Fred Smye tapped Dixon's shoulder. "What the devil," he whispered, "is a 'Super' Jetliner?"

Stan Dixon replied with a wry grin, "A Super Jetliner is our Jetliner with CPA markings."

* * *

In a winner-take-all contest worth a potential $12.7-billion, American Airlines named the Avro C804 as its new trans-Pacific twin-jet over the Boeing 767–500. The deal breaks down to twenty-eight firm orders and six options. American will also take possession of two options relinquished by the Canadian charter airline Wardair following that carrier's bankruptcy.

"On a plane-to-plane basis, Avro's C804 twin-jets were more expensive than [the 767]," American president Donald J. Carty said last week. "But price was not the only consideration." Carty cited technical specifications and an evaluation of overall costs as the most important factors in favour of the C804. Cabin size was also a dominant issue. "The C804 is about ten percent larger, so it has a revenue earning advantage," Carty said.

American's announcement ends a long and occasionally bitter marketing battle between Avro and arch-rival Boeing. Officials of the Seattle-based company could not be reached for comment. With the American order, Avro now holds commitments for 103 C804s. Other customers include JAL, United, Korean Air Lines, Delta, Cathay Pacific, Air Canada, United, CP Air, Qantas, and British Airways.

—Aviation Week and Space Technology, 2 July 1990

Afterword

The world's first commercial jet airliner was the de Havilland Comet, which made its first flight in July 1949 and entered passenger service with BOAC (now British Airways) in May 1952. In a bold technological leap, it seemed the British would leave the Americans behind with their rival Boeing 707 not entering service until 1958. But the Comet went on to suffer a series of accidents. Two crashed during take-off, the cause attributed to pilots inexperienced with jets, and a third went down in a thunderstorm. Later, two more Comets disintegrated in mid-air over the Mediterranean, killing all passengers and crew. An investigation led by the Royal Aeronautical Establishment concluded the design of the Comet's pressurized cabin had not adequately addressed metal fatigue, a structural failure mode that was not well understood at the time of the project's inception. The British aircraft industry never fully recovered, with the hard-won lessons of the Comet benefitting American companies that would dominate the market for decades until the rise of Airbus.

What is little known is that the world's second jet airliner was built in Canada.

The Canadian aeronautical industry owes much to the British. During the Second World War, a company called Victory Aircraft Ltd. manufactured British-designed bombers and transports under licence at a factory in Malton, Ontario near the present site of Toronto Pearson International Airport. Victory Aircraft ceased operations in 1945 and was sold to the British company Hawker Siddeley, which merged it into their Canadian subsidiary A. V. Roe Canada Ltd. (also known as Avro Canada).

James Floyd, a British-born Canadian engineer, was the chief designer of the C102 Jetliner. Responding to a specification from the government-owned Trans-Canada Airlines (TCA, now Air Canada), Floyd and his team at Avro created an aeronautical marvel. Powered by four Rolls Royce Derwent turbojet engines, the Jetliner could carry up to 50 passengers at a cruising speed of 720 km/hour over a distance of up to 1,700 km. The prototype Jetliner made its first flight on 10 August 1949, just thirteen days after the Comet. Over a series of demonstration flights in 1950, the Jetliner nearly halved the travel time on routes such as Toronto-New York, Toronto-Winnipeg, and Toronto-Chicago. Floyd's superb book *The Avro Canada C102 Jetliner* (Boston Mills Press, 1986) was a major source of research for "The Coming Age of the Jet".

TCA proved to be a difficult customer. Their contract to Avro was fixed price, but the airline kept asking for design changes that drove up the cost.

TCA eventually decided it did not want the risk of being the first airline to operate jets. Not only did they refuse to buy the Jetliner, but officials made negative public statements about the aeroplane in an effort to justify their decision. Nevertheless, Avro managed to secure expressions of interest from a number of U.S. airlines including United, American, Eastern, National, and TWA. Howard Hughes was so enthusiastic about the Jetliner he tried to arrange for it to be produced under licence in the U.S. for TWA. But all this was for naught, and in late 1950 the Canadian government ordered Avro to halt the Jetliner project and focus on the CF-100, a subsonic jet fighter for the Royal Canadian Air Force. In his book, James Floyd lamented that the Americans were more supportive of the Jetliner than his fellow Canadians.

But there was another Canadian who might have been interested.

Grant McConachie was born in 1909, the son of a railroad engineer. Much of McConachie's early career was spent as a bush pilot flying aircraft in the service of isolated communities in the remote northern regions of Canada. In 1942, the Canadian Pacific company bought McConachie's airline and merged it with nine other bush carriers to form Canadian Pacific Airlines. He rose quickly through the ranks and was appointed to the presidency in 1947. Grant McConachie's remarkable life and career are documented in Ronald Keith's biography *Bush Pilot with a Briefcase* (Douglas and McIntyre, 1998).

In my research for "The Coming Age of the Jet", I found no evidence that Avro ever contacted McConachie about the Jetliner. This seems a curious oversight on the part of Avro because McConachie was a technophile who was always looking for ways to get Canadian Pacific ahead of its arch-rival TCA (he really did envision suborbital rocket planes). After seeing the Comet at the 1949 Farnborough Air Show, McConachie was determined to make Canadian Pacific one of the first airlines in the world to operate jets and ordered two Comets for service between Sydney and Honolulu. On 3 March 1953, the first Canadian Pacific Comet crashed during take-off from Karachi, Pakistan for what was to have been the next leg of its ferry flight from London to Sydney. Canadian Pacific cancelled its order for the second Comet and would not operate jets until 1961 with the U.S.-built Douglas DC-8.

Would Grant McConachie really have been interested in the Jetliner? His lifelong passion was to fly from Canada to destinations in the Asia–Pacific. In 1949, McConachie met General Douglas MacArthur in Tokyo and secured landing rights in Japan. Shortly after the outbreak of the Korean War, Canadian Pacific was contracted to airlift American military personnel to Tokyo for service in Korea, ferrying more than 39,000 between August 1950 and March 1955. The modest range of the Jetliner would have been problematic for flights to Tokyo, but that could have been addressed with additional fuel tanks, new engines, and a series of fuel stops across the Aleutians and the northern islands of Japan.

In the real world, the Avro Jetliner was never put into production and the sole prototype was scrapped in 1956. Avro Canada would go on to develop a supersonic interceptor called the CF-105 Arrow that was itself cancelled in 1959. Lacking a diversified portfolio of customers other than the Canadian government, Avro Canada went out of business shortly afterwards. James Floyd returned to England where he worked on early engineering studies for what would later become the supersonic Concorde. Many other former Avro engineers emigrated to America where they would make crucial contributions to the U.S. space programme including the Apollo Moon landings. Grant McConachie suffered a fatal heart attack in 1965 while on a business trip doing what he loved most—taking delivery of a new aeroplane. Trans-Canada Airlines was renamed Air Canada that year but would remain a government carrier until its privatization in 1988. And in 1965 my father Douglas Choi would begin a long career as a proud employee of the airline that Grant McConachie built, retiring in 2001 shortly after Air Canada acquired its long-time rival.

Much of the history of aviation has been motivated by the quest for speed, from the propeller-driven DC-3 to the early jets like the Comet and the Jetliner to the supersonic Concorde. Today, aviation must be motivated by environmental sustainability. According to the U.S. National Oceanic and Atmospheric Administration, aviation is currently responsible for about 3.5% of all drivers of anthropomorphic climate change. Fuel efficiency could be improved with new engines, lighter materials, and low-drag aerodynamics such as very high aspect ratio wings and blended wing bodies. Future aircraft could be powered with cleanly sourced electricity or non-food crop biofuels. According to the Intergovernmental Panel on Climate Change, improvements in air traffic management (i.e. optimised flight paths) could reduce fuel burn by up to 18%. If mass air travel is to continue, there must be a coming age of the sustainable jet that we can all live with.

Crimson Sky

Press Release
Date: Ls 117.43, 59 A.L.
Source: The Bessie Coleman Foundation

A Voliris 3600 lighter-than-air vehicle took off today from Yeager Base, Arabia Terra, at 07:22 Coordinated Mars Time, launching a bold attempt to set a new Martian record for the longest flight made by an aircraft. Piloted by Carl Gablenz, with funding from the Bessie Coleman Foundation and support from Thomas Mutch University, the blimp is expected to fly over 600 nautical miles in approximately 80 hours. Gablenz is scheduled to land at Laurel Clark Station on the western edge of Isidis Planitia.

Link here for video and images of the departure.

* * *

Every med-pilot does their own things before flying.

If anyone were to ask about their routines, Martian med-pilots would swear that whatever they did was based on method and procedure, never superstition. Some of them, usually the grizzled veterans, hung out in the ready room, perhaps drinking coffee or watching videos or playing solitaire. Newbies might be found in the map room studying the latest mission profiles, or going over operational procedures in a simulator.

When she wasn't strength training in the gym, Maggie McConachie drank coffee and read journals while listening to the irregular beat of heliocentric jazz. Helio had been all the rage when she was growing up. Her dad had loved it, and she too had learned to relish its strange rhythms. She now read her journals to its siren call. Never aviation or medical journals, though— Maggie's pleasure reading was scientific journals. Dad had still been a grad student when she was a baby, and he would often lull her to sleep by singing papers he had to read, thereby killing two birds with one stone. Maggie might very well be the only person in the Solar System to find soothing comfort in the bizarre medley of heliocentric jazz and partial differential equations.

A framed still image of Maggie as a young child, with her father at her side, broke the grey monotony of the otherwise spartan walls of her quarters. Her dad used to travel frequently to scientific conferences and would often bring his young family along. Maggie must have been around two or three Earth years old at the time the picture was taken, in a boarding gate waiting area at the old LaGuardia Airport. They were standing in front of the windows looking out onto the apron, her father kneeling beside her as she pointed a short, podgy finger at a passing aeroplane.

The call came in at 08:41 MTC. Maggie was next up in the flight rotation. "Med-Three here."

The message was terse. She nodded and put down the reader, stealing a quick glance at the picture before dashing out of her quarters, the music fading to silence before the door closed behind her.

Navigating the claustrophobic hallways and ladders of Syrtis Station, she found her way to the operations centre in less than a minute. Ops was crowded, as usual, with teams of technicians seated at their workstations. Liu Huang, the Air Search Coordinator, turned to her and nodded as she entered the room. In the middle, surrounded by banks of screens, was Charles Voisin, the chief Search Master for the Mars Search and Rescue Service at Syrtis Station. Maggie approached Charles, carefully squeezing through narrow rows of equipment and workstations.

"Good morning, Maggie. I have an excellent mission for you." Charles was a slight man of medium height. His angular face was crowned by curly dark hair, with a neatly trimmed moustache and large soulful eyes that always had slight bags under them, as if he never quite got enough sleep. Maggie thought Charles looked a little bit like her dad when he was young. "We have an aircraft down."

"Where's the ELT?" Maggie asked.

Liu uploaded a panoramic map to the wall screen. A flashing icon with the registration M4-LGA indicated the approximate signal source from the downed aircraft's emergency locator transmitter. "Arabia Terra, near the southwest rim of Antoniadi Crater."

"That's getting awfully close to the bingo fuel radius of the chopper," Maggie said, referring to the farthest distance she could safely fly before having to either return to base or find an alternate landing site for fuel. The latter were extremely rare on Mars.

"There aren't any permanent settlements at Antoniadi yet. Who's out there?" Maggie paused for a moment. "Oh, for the love of…It's that guy trying to set the record, isn't it? Carl…Gablenz?"

"Yes."

"But he's only been up since…what, yesterday, and he's in trouble already? As if we're not busy enough already without having to pull damn stunt pilots out of their self-inflicted messes." Maggie made a face. "Isn't he supposed to be rich? Can we send this playboy the bill?"

"We do not go after people for costs just because they have the money to pay for it," Charles said gently. "Someone gets lost or goes down, we go help them. That's our job."

"Who says universal healthcare is dead, huh?"

Charles shrugged.

"All right, then. Liu, get me the METARs and PIREPs," Maggie said, referring to the meteorological aviation reports and pilot weather reports. "Start with the upper level weather – wind speed, bearing aloft, and temperature. I'll also need the forecasts and updates for the target area as well as current weather on-scene, especially site visibility."

"Roger that," said Liu. He called up a display. "We have a low pressure trough approaching the crash site from the northwest."

As Liu continued with the weather briefing, Maggie pulled out a tablet to prepare her flight plan.

"We have requested Mr. Gablenz's medical records from Earth," said Charles. He consulted another display. "The Harmakhis-7 satellite will be passing over that area in about twenty minutes. We will transmit all data to you en route as it becomes available."

"All right, Charles." She pronounced his name Anglo style, with a hard "ch" sound.

"Soyez prudent, Maggie."

She looked at him with a blank expression.

"You have no idea what I just said, do you?" His moustache twitched in amusement. "No matter, although I wish you would at least try to pronounce my name correctly."

Maggie tapped the tablet to file her completed flight plan. "Just make sure the coffee's hot when I'm back." She dashed out of the operations centre and went to put on her biosuit. Ten minutes later, she was on the pad.

MarsSAR employed the Bell-Xīnshìjiè BX-719A helicopter. A two-armed dexterous robot nicknamed Chop-Chop performed near-continuous systems diagnostics and routine line maintenance for the BX-719 on ready standby. The waiting vehicle was further checked every couple of hours by a human technician who performed a more detailed inspection and then signed-off the helicopter as ready to fly. This minimised the time between a call coming in and when the med-pilot could be dispatched.

"Liu, please confirm the flight status of vehicle," Maggie radioed.

"*The last A-check was completed 38 hours ago,*" Liu reported. "*No major faults. One minor fault, an intermittent indication on the starboard landing light status, not a MEL issue. Caution memory is clear. Vehicle flight status is green.*"

"Thanks, Liu."

Formal assurances aside, Maggie always made a point of taking a minute to do a quick check herself. After one of her early flights, a technician on the Air Search Coordinator's team—perhaps insulted by her apparent lack of trust—asked her why she did it. She told her the truth: "Because I want to stay alive." Chop-Chop took no offence.

Every med-pilot does their own things before flying.

Jumping into the seat, Maggie checked the status of the liquid hydrogen and oxygen tanks, the regenerative fuel cells and the batteries, as well as the on-board medical equipment. Finding everything in order, she hopped out and did a quick circuit around the chopper, starting from the port side and working counter-clockwise. On the ground, the BX-719 sat on four landing legs with articulated foot pads. Maggie looked for leaks in the shock absorbers of the portside pair. She then scanned the port engine pod and its ten-bladed propeller for damage. The BX-719 was equipped with pusher props on each side, which served to increase the chopper's speed while countering the torque of the large main rotor through differential thrust.

She then climbed to the top of the helicopter and looked at the transmission well and the main rotor for anomalies. The BX-719's rotor had four low aspect ratio blades made of reinforced Kevlar epoxy skin stretched over a skeleton of graphite epoxy spars and ribs. Resembling giant fan blades, they were twisted along their lengths, and the top and bottom surfaces were equipped with a pair of upper and lower boundary layer trips to produce an optimal lift distribution.

After jumping down, Maggie went to look at the last major component of the helicopter, a large V-shaped horizontal stabilizer at the rear of the aircraft. She scanned the elevator and trim tab, and then manually moved the elevator up and down. Once the portside check was finished, she repeated the procedure on the other side of the helicopter.

With her personal inspection ritual completed, she returned to the cockpit, strapped herself into the pilot's seat and plugged her biosuit into the helicopter's power and life support system. With the exception of a large forward windshield, the cockpit was open and unpressurized. She powered up the flight management system and avionics, started the engines, and commenced the take-off procedure.

"Syrtis Station, MarsSAR-3 is ready for departure."

"*MarsSAR-3, you are cleared for departure. Surface winds are from two-seven-zero at eleven knots, gusting to twenty. Good luck, Maggie.*"

Maggie confirmed the callout with the meteorological data displayed on the augmented reality projection on the inside of her helmet. "Thanks, Liu."

She raised the collective with practiced confidence and brought the helicopter to a hover over the pad. After a final check of the instruments and the flight controls, she applied more collective and pushed the cyclic forward, translating the BX-719 to forward motion.

This was already Maggie's sixth mission since being assigned to Syrtis less than eight Earth months ago. By necessity, they were all solo missions. A lone med-pilot plus the patient (or two, if the latter were light enough) was all the helicopter could lift in the thin Martian atmosphere. If there were more casualties, she could only take back the one or two most critically injured. For the remainder, she would do her best to stabilize them on site, to await either her return or the arrival of a MarsSAR ground team.

Every mission was different, but there were also similarities—most notably, the way she felt during the outbound flight. Like many young pilots, she was always geared up, her adrenaline constantly pumping. She knew exactly what she had to do; her training made that a certainty. Yet, at the back of her mind, there were always questions: *How am I going to pull this off? What surprises await me?*

Maggie didn't know much about this Carl Gablenz character, just brief clips of stuff she'd seen on media. He was probably one of those self-made rich people who had racked up a fortune in finance at Clavius. Somebody once tried to climb the four "Mons of Tharsis" in one year but quit after getting stuck somewhere halfway up Pavonis. Maggie thought it might have been Carl. She was pretty sure he was the guy who had tried to do a solo balloon circumnavigation of Titan. That had been a failure as well. Maggie wondered if he'd ever succeeded in any of his stunts.

If nothing else, she really hoped he was still alive when she found him. The paperwork for processing dead people was horrendous.

Maggie's thoughts were interrupted by a radio report from Liu.

"*I have good news and bad news,*" he said. "*Which would you like first?*"

"Surprise me."

"*Here is the data from Harmakhis-7, hot off the downlink.*" As he spoke, an image appeared on Maggie's augmented reality display showing a grey truncated ellipsoid with stubby fins against a crimson background. "*We have pinpointed the crash site, and the coordinates are being entered into your FMS now.*"

Maggie confirmed that the helicopter's flight management system had accepted the navigational data. "I take it that's the good news. What's the bad?"

Another image appeared inside her helmet. At first, it appeared to show a featureless Martian plain. But as the contrast was enhanced, a pair of lines cutting across the surface became visible. They resembled shallow trenches, somewhat like those left by fingers scraping across fine sand, but on a much larger scale. According to the display, they were several hundred metres in length.

"Dust devil tracks," she said grimly.

"*Yes. They are probably what brought down our intrepid adventurer Gablenz.*" If it had been at low altitude, the slow-moving blimp and its possibly tired pilot would have been easy prey for the strong whirlwinds.

Maggie gritted her teeth. "So which Department bureaucrat should we call to ask about our lidars?" The MarsSAR fleet was supposed to have been equipped with laser detection and ranging units months ago. Remote Syrtis Station was still waiting.

"*Be careful, Maggie. Syrtis Station out.*"

She frowned, contemplating her situation. Martian dust devils were difficult to see, and without a lidar system there was no reliable way to detect one until she literally flew into it. But she remembered reading a journal paper about how the swirling dust often became charged through triboelectric effects, producing low frequency radio emissions. Maggie tuned one of the helicopter's receivers to pick up in the lower AM band. She wished she had more data, but with luck the radio might give her a few seconds of warning.

Maggie let the autopilot fly most of the course, guided by data from the Harmakhis-7 satellite. She took over manual control as she approached the crash site, flying a circular observation run around the downed aircraft.

"I have a visual of the target," Maggie reported. "Video and data telemetry on. Attempting to link-in with the aircraft's flight data recorder." The link status icon on her augmented reality display remained a red X. "No joy. Liu, where are we?"

"*A-OK on your data and video, I'm seeing you fine. No link to the FDR. Please try again.*" Liu's voice crackled over the radio. "*We... picking up interference...*"

"Copy that," Maggie replied.

Maggie continued to circle the crash site, transmitting video and data back to Syrtis. The blimp was tilted about thirty degrees to the surface, its cruciform fins pointed in the air and its crumpled nose planted into the ground. Except for the ruptured forward ballonets, which had lost their hydrogen harmlessly to the carbon dioxide Martian atmosphere, the solar cell-covered envelope still largely retained its shape. The left-side ducted-fan thruster pod was damaged, but otherwise the gondola housing the pilot also appeared relatively intact.

"*...doesn't look good,*" Liu said.

"No, it doesn't."

"*Still...no link.*" Liu's voice was still dropping out intermittently. "*Their communications subsystem...damaged, proceed...caution...*"

"Boys and their toys," Maggie muttered. "Why do we let idiots do these stunts?"

Maggie landed about a hundred feet from the crashed blimp. As the helicopter's huge fan-like rotors slowed, she released her seat harness and switched life support from the helicopter's to her biosuit's internal system before disconnecting the umbilical and climbing out of the cockpit. Maggie went around to the helicopter's trauma bay and deployed the stretcher, picking up the medical kit and portable life support unit before making her way out to the crashed blimp. It was a physically demanding task, even in three-eighths gravity.

"*Syrtis Station to MarsSAR-3.*" This time, the radio was clear.

"Go ahead, Liu."

"*We have received Mr. Gablenz's medical records from Earth.*"

"Any allergies or relevant preconditions I should know about?"

"*None.*"

When Maggie got to the unpressurized gondola, she found Carl Gablenz unconscious, still strapped in his seat. Carl's biosuit, like Maggie's, was a sleek, form-fitting garment that applied counter-pressure to the body mechanically rather than barometrically with air like the bulky old spacesuits of the first human Mars landings. Maggie peered into the hard, transparent bulbous helmet. Carl looked younger than the twenty-five Martian years indicated in his medical records. With his eyes shut, his roundish face looked almost serene, and his black hair had only the slightest streaks of grey. She could not see any obvious signs of an airway obstruction like vomit, and a small patch of condensate on the inside of the helmet showed he was still breathing. Carl was indeed alive—to Maggie's great relief.

With efficient skill, Maggie unplugged Carl's biosuit umbilical from the blimp's dying life support and connected it to her portable unit. She initiated a wireless link with the biosuit computer and transmitted the MarsSAR key to access the embedded medical sensors. Next, she commanded the biosuit's smartskin to rigidize in order to immobilize its occupant as much as possible. On Earth, or in a pressurized Martian habitat, Maggie would have checked her patient's blood circulation by pressing their finger or toe nails and observing the capillary refill, but this was not possible through biosuit's gloves.

"Syrtis Station, this is MarsSAR-3. The patient is unconscious but breathing. Biosuit integrity has not been breached. His mean arterial pressure is sixty-seven."

"*Copy that,*" said Liu.

Suddenly, Carl let out a low moan.

"My name is Maggie McConachie, from Mars Search and Rescue," she responded calmly. "You're going to be fine. We'll get you out of here very soon."

She would soon have to move Carl, but there was nothing more she could do to restrain his neck and cervical spine beyond rigidizing the biosuit's smartskin. Attempting to insert a brace or splint would require taking off his helmet. EVA trauma protocols still left much to be desired. It was medical heresy to not better restrain the neck, but she had no choice but to be careful and keep any necessary motions as gentle as possible.

"Syrtis Station, the patient is semi-conscious," Maggie reported. "Pulse steady, blood pressure systolic 80, respiratory rate 12, temperature 37.6. Level of consciousness is GCS 5. I'm going to oxygenate him now." She commanded the portable life support unit to vent the air in Carl's helmet and replace it with pure oxygen. She could see his eyes start to flutter. He looked like he was trying to say something. Maggie felt Carl's legs and arms, looking for signs of broken bones and finding none. "I'm going to administer Ringer's lactate for fluid volume resuscitation."

"*Data...*"

Maggie blinked. The voice on the radio was not Liu's. "Mr. Gablenz?"

"*Important, data...*"

"Don't worry, Mr. Gablenz," Maggie said. "We'll have you on your way very shortly. Everything's fine." She pulled an EVA syringe from her med kit and jabbed it hard into Carl's left forearm. The Ringer's solution was delivered in seconds, and Maggie withdrew the syringe. A normally functioning biosuit's smartskin could self-seal millimetre-sized punctures, but for the sake of time Maggie simply slapped a patch over the pinprick.

"*We have a yellow caution on oxygen constraints,*" Liu warned. "*You'd better start heading back to the chopper soon.*"

Maggie pulled the stretcher up beside Carl. She was about to release the harness that held him in the pilot's seat when she noticed a small still image stuck to the control panel. It was of a young girl, probably about one or two Mars years of age, sitting in the flight deck of some aircraft or spacecraft, pointing at the displays and controls.

Carefully supporting the upper body to minimise neck movements, Maggie slowly slid Carl off the seat and onto the stretcher. She briefly derigidized the biosuit's waist to lay him down, relocking the smartskin after he was fully reclined and strapped in. With the patient secure, Maggie began to push the stretcher back to the helicopter. She had just pulled up to the trauma bay when suddenly the stretcher began to thrash ever so slightly.

"*Get...data...*"

"Data?" Maggie repeated. She thought about the blimp's flight data recorder and her earlier inability to link-in. But there was nothing more she could do now. The recorders would have to be physically recovered whenever the crash investigation team from the Mars Transportation Safety Board showed up.

"Sir, I cannot recover the FDR data at this time," Maggie explained. "That will have to wait for the MTSB team. There is no time to go back to the wreckage now."

"*Not flight data...science...*"

"What?"

"*...data chit, my cuff...*" Carl lapsed back into unconsciousness.

Maggie looked at Carl's arms and spotted a small velcro-sealed pocket on the biosuit near his left wrist. Her finger fished inside and produced a data chit. She stared at the small sliver for a moment before putting it into her own biosuit's pocket. Then she docked the stretcher to the helicopter's trauma bay, deflating the wheels before pushing it all the way inside and securing it. Finally, she unplugged Carl's biosuit umbilical from the portable life support unit and connected it to the helicopter's system.

"Syrtis Station, this is MarsSAR-3," Maggie said as she strapped herself back into the pilot's seat. "The patient is secure. I am commencing my return now."

Maggie raised the collective and the helicopter lifted off from Arabia Terra, kicking up a small amount of ruddy dust in its wake, and headed in a south-easterly direction back towards Syrtis. Maggie watched the altimeter display on her augmented reality visor count up past 1,000 feet above ground level.

She activated the autopilot and settled back in the seat, occasionally glancing at the display in her helmet that was monitoring Carl's medical parameters such as heart rate, body temperature, respiration, blood pressure, and oxygen saturation. Her thoughts turned to Carl's data chit. She pulled it out of her pocket and plugged it in. Another display popped into her helmet, showing parameters of a different kind: wind speed and direction, temperature, barometric pressure, relative humidity, atmospheric opacity.

"Meteorology." At last, she understood. "Carl was collecting science data."

"*Maggie,*" Liu called in, "*when you have a moment could you please transmit –*"

"Hey, Liu? I didn't copy your last –"
The AM radio crackled to life.
"Tabarnak!" Maggie immediately disengaged the autopilot, pushing the cyclic forward and pulling hard on the collective. The helicopter began to accelerate, and the altimeter reading crept past 2,000 feet.

A few seconds later, it hit.

Maggie was pressed into her seat as the helicopter abruptly lofted upward. A moment later, she felt the seat drop out from under her, and she was slammed hard against the harness. She struggled to compensate as the helicopter yawed violently to starboard, but the controls were sluggish. On her augmented reality display, every icon that had anything to do with the helicopter's electrical system was lit up. She lost the flight management system and the avionics, and the fuel cells went offline.

With painful slowness, the controls began to respond. Maggie managed to stop the yaw and levelled out the helicopter. The buffeting subsided, and she felt the BX-719 climbing again. With the flight management system out, she could only guess at how much altitude she had lost, but one look down told her it had been very close. She could see individual rocks on the surface.

"*– respond please, Maggie. Are you all right?*"

"Liu! Yeah I, uh – I think I'm still alive. Gimme a second here." She switched to the backup flight management system and power cycled the avionics. Live data began to reappear on her augmented reality display.

Another voice came on the radio. "*Maggie, this is Charles. I am happy to learn that you do speak a sort of French after all, but I would advise you not to say such things in polite company. What happened?*"

"I just made the acquaintance of the devil." Maggie blinked, trying to clear the sweat that had run into her eyes. "Nearly ran me into the ground, and the electrostatic discharge fried a bunch of stuff. I'm running on the backup FMS and the batteries. Wait a minute –"

A status icon changed.

"OK, the fuel cells have reset and are back online." She checked the medical telemetry. "And our guest is OK. Slept through the whole thing, so to speak."

"*Do you need assistance?*"

"Yes, I need assistance...make sure the coffee's hot when I get back!"

"*Copy that, Maggie,*" said Charles. "*You have certainly earned it.*"

She levelled out the helicopter at an altitude of 5,000 feet above datum for the flight back to Syrtis. The late morning Sun cast a diffuse light over the endless bloody plains below her, a landscape wounded by craters and smothered by a crimson sky.

Maggie's thoughts turned to Carl Gablenz.

On the Earth of the past, it was the pilots who had blazed the trails into the frontiers of the day. Over continents and oceans and across the globe, there was always someone who had to do it first so that others could follow. Flying started as adventure for the few, and through their daring eventually became a safe and indispensable means of travel for all. As it was on the blue planet, so it is again on the red.

Carl Gablenz was not a stuntman. He was a pioneer, and somewhere another small future explorer was waiting for his safe return. Perhaps the two of them really weren't so different after all. They might even do the same things before flying.

"Coffee." Maggie McConachie smiled. An atmospheric physics journal and some heliocentric jazz, she decided, would go very nicely with that.

*   *   *

Press Release
Date: Ls 118.74, 59 A.L.
Source: The Bessie Coleman Foundation

Carl Gablenz has been rescued by the Mars Search and Rescue Service and is currently recovering at the Syrtis Station medical facilities. Mr. Gablenz expressed his deep gratitude to the courageous personnel of MarsSAR, and thanked all those who have sent well-wishes from across the Solar System. Although his record-setting flight attempt was cut short, valuable scientific data was collected that will help researchers at Thomas Mutch University improve their models of the

Martian atmosphere, which promises to make future air travel safer. Mr. Gablenz also vowed to make another attempt at the Martian flight duration record as soon as possible.

"It's all part of the process of exploration and discovery," said Mr. Gablenz. "It's all part of taking a chance and expanding our horizons. The future doesn't belong to the fainthearted; it belongs to the brave."

Afterword

"Crimson Sky" was first published in the July/August 2014 issue of *Analog Science Fiction and Fact*. Less than seven years later, on 19 April 2021, a small robotic Mars helicopter appropriately named Ingenuity made the first powered and controlled flight of a heavier-than-air craft on another planet. Developed by a team at the NASA Jet Propulsion Laboratory under the leadership of project manager MiMi Aung and chief engineer Bob Balaram, the Ingenuity helicopter on its first flight climbed to an altitude of three metres and maintained a stable hover for thirty seconds before a controlled return to the surface of Mars. Aboard Ingenuity was a piece of wing cloth from the Wright Brothers' Flyer, which made the first successful powered and controlled flight of a terrestrial heavier-than-air craft on 17 December 1903—a symbolic linking of aviation milestones across 117 years and two worlds.

Designing an aircraft to fly on Mars is very challenging due to the nature of the Martian environment. The lower Martian gravity means that for the same mass an aircraft would only weigh 38% as much as it does on Earth, however, the atmosphere is also a hundred times less dense. At the Martian surface, the air density is equivalent to the Earth's at an altitude of thirty kilometres, so an aerofoil would need approximately five times more surface area to generate the same amount of lift. In addition, the low ambient temperature and low atmospheric pressure makes the local speed of sound lower on Mars than on Earth, meaning that transonic aerodynamic effects such as shock waves are encountered at lower flight velocities. This is particularly challenging for a Mars helicopter like Ingenuity or the fictional BX-719A because the forward-sweeping (leading) edge of the rotor blades (or the tips of propellers) could approach or exceed the local speed of sound even if the rest of the aircraft does not. Special care must be taken in the aerodynamic design to ensure that an adequate lift-to-drag ratio is maintained over the expected flight conditions.

The first human to fly on Earth was Jacques-Étienne Montgolfier, who in October 1783 lifted off in a hot-air balloon he had designed with his brother

Joseph-Michel Montgolfier. Solar-heated balloons of the Montgolfier type have been proposed for Mars exploration.[1] Such a balloon would be made of a heat absorbent material with an open bottom to be filled with carbon dioxide from the ambient atmosphere. In the morning, the Sun would heat the gas inside the balloon until there is sufficient buoyancy to rise and be carried aloft by the winds. At night, the balloon would descend with the cooling temperatures and arrive at a different part of the planet. A small closed helium balloon would be required to keep the main hot-air balloon upright and prevent it from getting stuck on the ground. The balloon could either carry a gondola with scientific instruments, or in one concept, a snake-like tether that would allow instruments to touch the ground at night. A prototype of the latter was successfully tested in the early 1990s by (appropriately) the French space agency CNES in partnership with Russian and American scientists.[2] Dirigibles like the fictional Voliris 3600 have also been proposed, but so far no lighter-than-air vehicle has yet been flown on Mars.

The fictional BX-719A combined elements from two Mars aircraft designs, a helicopter concept called MARV (Martian Autonomous Rotary-wing Vehicle)[3] developed by a team at the University of Maryland led by Anubhav Datta and Inderjit Chopra, and a fixed-wing biplane concept called Zephyr[4] developed by Brian Morrissette and James DeLaurier at the University of Toronto. The Ingenuity helicopter bears some resemblance to MARV in that both employ coaxial rotors where two sets of blades spin in opposite directions to mutually cancel out their torques. In contrast, the fictional BX-719A was equipped with pusher props on either side of the fuselage that counteracted the torque of a single large rotor through differential thrust. One of the challenges in writing "Crimson Sky" was how to describe the design of the helicopter without resorting to the notorious "core dump" of information. This was solved by making the description a part of the pre-flight walk-around inspection performed by the Maggie McConachie character. Retired Major Jonathan Knaul of the Royal Canadian Air Force, a pilot who flew helicopter missions in Kosovo and Afghanistan, was an invaluable resource in helping me understand the real-life ground and flight operations of these aircraft.

---

[1] Blamont, Jacques. "Exploring Mars by Balloon", *The Planetary Report*, Volume VII, Number 3, May/June 1987, pp. 8–10.

[2] Anderson, Charlene M. "Wind, Sand and Mars: The 1990 Test of the Mars Balloon and SNAKE", *The Planetary Report*, Volume XI, Number 1, January/February 1991, pp. 12–15.

[3] Dhatta, A. et al. "The Martian Autonomous Rotary-wing Vehicle (MARV)", Department of Aerospace Engineering, University of Maryland, 1 June 2000.

[4] Morrissette, B. R. and DeLaurier, J. D. "The Zephyr: Manned Martian Aircraft", *Canadian Aeronautics and Space Journal*, Volume 45, Number 1, March 1999, pp. 25–31.

The biosuits worn by the characters in "Crimson Sky" were based on designs developed by Dava Newman's team at the Massachusetts Institute of Technology.[5] Unlike the familiar traditional bulky spacesuits, the biosuit is a sleek form-fitting garment that applies counter-pressure to the human body mechanically rather than barometrically with air. Wearable technologies such as the medical sensors described in the story could be embedded within its layers. Such a suit would also be light and flexible enough to allow astronauts to easily and comfortably walk, run, or even scale mountains—perhaps what the Carl Gablenz character was trying to do when he got stuck on the slopes of Pavonis Mons.

References and homages to aviation and space history abound in the story. Bessie Coleman was the first Black woman to earn a pilot's licence in the United States. Yeager Base is named for Chuck Yeager, the U.S. Air Force pilot who broke the "sound barrier" in 1947. Thomas Mutch University is named for the American planetary scientist who (if you remember from "Dedication") is also memorialized at the Viking 1 landing site on Mars. Laurel Clark Station is named for the NASA astronaut who was killed in the Space Shuttle *Columbia* accident in 2003. She appears in the alternate history story "A Sky and a Heaven". Finally, the protagonist Maggie McConachie is implied to be a descendent of the Canadian aviation pioneer Grant McConachie, the hero of "The Coming Age of the Jet".

"Crimson Sky" was the recipient of the 2015 Prix Aurora Award, Canada's national award for excellence in speculative fiction, in the category of Best Short Fiction English.

---

[5] Heger, Monica. "What to Wear on Mars", *IEEE Spectrum*, June 2009, Page 34.

Most Valuable Player

Not long ago, Terry Jospin thought he'd never carve a turkey again. Though it looked lifelike, functionally his prosthetic right hand was little better than a claw. It had been over a year since the accident, but Terry was still struggling to perform everyday tasks.

"I'm sorry, but I said I wanted dark meat."

"*What?*"

"Dark meat," said Daniel Hanley quietly.

Terry threw down the knife. "I give up."

Marie Jospin started to say something.

"Don't start, Marie! Just because he's your brother –"

"I'm full, Daddy," interrupted five-year-old Mark Jospin.

"Finish it! Don't waste your food."

The meal continued in silence. Finally, Daniel spoke up. "Hey, how about those Tucson Tarantulas. They whipped the New York Yankees in seven games."

Terry looked down at his food, but Marie responded. "Yeah, that last game was amazing. Danny Daniels hit a home run for the Yankees, but Vic Caruso socked two dingers for Tucson to ice the game."

"I like Caruso, too. Can you believe he's fifty-three? It's those stem cell and telomerase treatments. Caruso's made the Old-Timer's League All-Star Team the last two years, hitting over .300 both seasons." Daniel continued, "Of course, even he wasn't as good as Ty Cobb. Did you know that *no* player has ever beaten Cobb as the greatest hitter in history?"

Terry stood. "I'm eating in the family room."

© The Author, under license to Springer Nature Switzerland AG 2016
E. Choi, *Just Like Being There*, Science and Fiction,
https://doi.org/10.1007/978-3-030-91605-3_7

Twenty minutes later, he heard the car start. Marie was taking Daniel to the bus station. He returned to the kitchen to clean up. By the time he finished, his wife had returned.

"Why do you keep inviting him here?" Terry asked. "He's twenty. He can take care of himself."

"It's Thanksgiving, and he's the only family I've got." She looked him in the eye. "Daniel's quite a baseball fan. He really looks up to you."

"He's a geek," he spat. Daniel Hanley was majoring in mathematics at UC Berkeley. Terry hadn't taken a math or science course since high school. "What the hell does he know about baseball?"

* * *

Terry fumbled the keys, but eventually managed to open the door. He tripped on the raised step, but kept his balance long enough to enter the house without falling.

"Hey," he called out, "where's dinner?"

Marie emerged from the shadows, standing in the frame of the kitchen door. "You stumble in at this hour, and all you can say is 'where's dinner'?"

"I was just out –"

"You were out with strangers, bragging about yesterday's glories and trying to drown today's sorrows."

"Those guys are my –"

"Friends? No, Terry. They're strangers."

"They *listen* to me, which is more than –"

"They don't listen, Terry. They don't care about you."

"Oh, *they* don't care. What the hell have *you* done –"

"How *dare* you say that?" Marie snapped. "It's now almost two years since the accident, and in all that time…I've tried so hard, but you just push me away. You push Mark away, and you treat Daniel like a freak –"

"No!" Terry exclaimed. "*I'm* the freak!" He raised his prosthetic hand. "I'm not even a whole man!"

Marie took a deep breath. "Terry, you're not a baseball player anymore. You'll never beat Ty Cobb's record. But you're still a father…and a husband. And I'm sorry, but you've been a miserable failure at both."

"What does *that* mean?"

Tears welled in her eyes. "I remember this guy who once missed the first day of spring training so I wouldn't have to go to that prenatal class alone. Have you seen him lately?"

* * *

Terry opened the freezer, retrieved a frozen dinner, and stuck it in the microwave. The kitchen, and indeed the rest of the house, was empty. Marie had won custody of Mark back in April. In the den, the TV was on. He wasn't watching, but he needed to hear a human voice in the background.

The microwave beeped. As Terry got up to retrieve the meal, his phone rang. He glanced at the screen, and caught his breath upon recognizing the number.

"Hello, Terry."

"Hi," he said. "So…how are you?"

"All right. Yourself?"

"OK." He paused. "It's good to hear from you, Marie."

"I know it's short notice, but…have you eaten yet?"

Terry glanced at the microwave. "No."

"Would you like to join Daniel and me for dinner?"

"I'll be right over."

Since the divorce, Marie and Mark had moved into a small apartment with Daniel. Terry felt nervous when he rang the doorbell. It was much as he'd felt on a night many years ago, when he first came to the home of a certain young woman.

"Come in, Terry. Dinner will be ready soon."

"Thanks again for having me."

"It's nothing," Marie said. "Listen, I have to pick up Mark. I'll only be a few minutes." She raised her voice. "Daniel! Terry's here."

Her brother emerged from a bedroom.

"Daniel," Marie said, "why don't you show Terry what you've been doing?"

At first, neither man moved. Daniel finally turned and went back inside. Terry followed.

The room was a shrine to baseball. Posters of the greats of past and present adorned the walls. An Oakland Athletics pennant hung over the bed.

Daniel turned on his computer and brought up a spreadsheet.

Terry sat on the bed. "Daniel, I never knew you were such a baseball fan."

"A lot of us are," Daniel said.

"Us?"

"Mathematicians. We love baseball. It has the most detailed numbers of any sport."

Figures filled the screen.

"What's this?" Terry asked.

"Something I've been working on. See, I've always wondered why yesterday's players always look better, statistically, than today's players. I mean, look at Ty Cobb. History records his batting average as .367. That record still stands."

Terry shrugged. "A batting average is a batting average: Number of hits over number of at-bats."

Daniel smiled. "That's *exactly* the problem. Baseball still calculates batting averages the same way it did in Cobb's day, and that's wrong."

"Why?"

"Look at it this way. A long time ago, shoes cost three dollars or something. Now, you can't get a good pair for less than, uh…"

"A lot more than three bucks," Terry said. "I see what you're saying. Just because the shoes were three bucks, it doesn't mean that things were cheaper back then. It's all to do with the economy and inflation and stuff like that."

"Exactly!" Daniel clicked the mouse, and ten columns appeared. "I've been looking at the stats for some of baseball's greatest hitters: Cobb, Rod Carew, Shoeless Joe, Roger Hornsby, José Vidro, Honus Wagner, Ted Williams, Danny Daniels, Stan Musial, Vic Caruso, Wade Boggs…" Daniel turned to Terry "…and you."

"What did you find?"

"It's like the price of things and the changing economy. Baseball's changed, so I've got to apply adjustments to fairly compare statistics from different years."

"First there's the league batting average. That adjusts for how easy or hard it was to get a hit in a given year. I'm taking into account things like lowering the pitcher's mound and expanding the strike zone, and I factor the emergence of night baseball when hitting is harder, and the increasing use of relief pitchers."

Daniel clicked the mouse, and the numbers changed.

"The second thing is ballpark effects. I figure Cobb's average was helped 6.4 points by favourable hitting conditions at Tiger Stadium, while yours was *hurt* 1.4 points by the layout of the Oakland Coliseum."

Another click, and the numbers changed again.

"Third thing is variability in talent among players of different eras. If you think about it, with baseball being a professional career and things like farm teams and spring training, today's players are much better than in Cobb's day."

"Cobb had it easier," Terry realized. "His competition wasn't as tough, so he looked better by comparison."

Click.

"Finally, to compare batting averages fairly, I limit comparison to the first eight thousand at-bats so that I'm comparing hitters still in their prime." Daniel clicked the mouse one more time. Terry leaned closer to the screen.

Ty Cobb's adjusted batting average was .342. His was .343.

Terry laughed. "So, I g-guess…I guess I really *am* the greatest hitter of all time!"

"Well," Daniel said, "I'm not sure everyone will agree with my math –"

"*Math!* Man, who'd have thought…"

"Hi, Daddy!"

He turned. Marie and Mark were standing at the door.

Terry got up off the bed slowly. "Why, Marie?" he whispered. "Why do you always know how to make me feel better?"

Marie gently ran her fingers across Terry Jospin's cheek. With profound sadness, she shook her head.

Afterword

Mathematicians and statisticians love baseball. It has the most detailed numbers of any sport. The nature of the game lends itself to record keeping and statistics: batting average, RBI (runs batted in), home runs, ERA (earned run average), strikeouts, stolen bases. By some accounts, there are more than a hundred statistics commonly and uncommonly used in baseball.

"Most Valuable Player" was inspired by a study in the late 1990s that was conducted by Michael Schell, a biostatistician at the Lineberger Comprehensive Cancer Centre at the University of North Carolina at Chapel Hill. Schell normally used statistical analyses to determine the efficacy and side effects of cancer treatments, but as a lifelong baseball fan he became curious as to why historical players look better statistically than modern players. Specifically, Schell was interested in why the legendary Ty Cobb, who played professionally from 1905 to 1928, still held almost seventy years later the record for the highest career batting average in American major league baseball.

A baseball player's batting average is calculated by dividing the numbers of hits with the number of at-bats. Schell argued that American professional baseball has changed drastically since Cobb's day, so batting averages should be adjusted statistically in order to more fairly compare players in different historical periods. He proposed modified batting averages with a number of factors: ballpark effects (how easy or difficult it is to get a hit in a given stadium), league batting averages (how easy or difficult it is to get a hit in a given year), and the variability in talent among players of different eras.

An important feature associated with ballpark effects is the size of foul territory. The bigger the foul territory of a given stadium, the more opportunity there would be for an opposing player to catch a batter out. League batting averages rose or fell in different years due to changes such as lowering the pitcher's mound and expanding the strike zone. According to Schell, a league average adjustment is also needed to account for the emergence of night baseball, when hitting is more difficult, and the increased use of relief pitchers.

Schell also believed it was important to adjust batting averages for differences in overall player talent between different eras. Ty Cobb, he argued, played at a time when his competition was not as strong and therefore he looked better in a relative sense. The development of major league baseball as a full-time professional career, particularly in the United States, and the emergence of farm teams and spring training meant that modern players are generally in better shape than those in Cobb's day. Finally, Schell argued that comparisons should be limited to the first 8,000 at-bats (using 4,000 as a minimum qualification) in order to assess hitters only in their prime and avoid favouring those who retired early over those who continued playing after age reduced their physical abilities.

"Most Valuable Player" was first published in the April 2016 issue of *Analog Science Fiction and Fact*. The fictional baseball players Vic Caruso and Danny Daniels, and the fictional team of the Tucson Tarantulas, were references to Ben Bova's light-hearted story "Old Timer's Game" in the anthology *Carbide Tipped Pens* (Tor, 2014) which explored the consequences of older professional athletes using stem cell and telomerase treatments to maintain their physical abilities and extend their careers into advanced age.

A Man's Place

Jamie Squires was dicing onions for an omelette when the alarm sounded.

The klaxon blared through the small confines of the kitchen, synchronized with the flashing red light on the ceiling. Jamie put down the knife, and after a quick check to ensure everything in the kitchen was off, ran out into the mess hall. The diners stood quickly from their seats, in some cases knocking chairs backwards. Their faces were apprehensive.

"*An X12 solar flare is in progress,*" said Laura Crenshaw, the general manager of Maryniak Base, over the intercom. "*All personnel are to report to their designated storm shelters immediately.*"

Billy Lu, Maryniak's chief engineer, appeared in the doorway. His red cap designated him as the emergency warden for this sector. "All right everyone, follow the signs, straight down the corridor. Let's move!"

Jamie followed the crowd into the passageway. He tried not to think about the X-rays and gamma rays that were even now going through their bodies. Traveling at the speed of light, they hit Maryniak at about same time as the warning from the space weather satellites at the L1 point. The imperative now was to get to the storm shelters before the arrival of the protons and heavy ions.

Joe McKay, the Shift 2 foreman, stood at the entrance to the shelter. "Right this way, people!" he said, pointing to the hatch in the floor.

Jamie mounted the ladder and lowered himself down the tunnel. Across Maryniak, personnel were gathering in six other protective chambers buried beneath the base's larger modules. The structures and the lunar regolith were supposed to protect the crew from the incoming stream of solar particles.

© The Author, under license to Springer Nature Switzerland AG 2003
E. Choi, *Just Like Being There*, Science and Fiction,
https://doi.org/10.1007/978-3-030-91605-3_8

There were already a dozen people in this shelter. Jamie found himself a spot on the bench along the wall. Ten more descended the ladder, followed by Joe and Billy.

"Is that it?" Joe asked.

"Crenshaw's on her way," Billy said.

The base manager arrived a few minutes later. "All set?"

Billy did a head count. "That's everyone for here."

"Close it up," Crenshaw ordered.

Joe climbed the ladder to close the outer hatch. Once he was back down, Billy slid the ladder up the tunnel before trying to close the inner hatch. The hinges creaked, and he seemed to be having difficulty engaging the latches, but he finally managed to seal the door.

"What's our status?" Joe asked Crenshaw.

"We're the last ones to lock down," she reported. "All personnel, both in-base and EVA, are in shelters. The proton stream should be sweeping through here in about twenty minutes."

"Are we sure this thing is buried deep enough?" Jamie asked nervously. He looked around, and was disappointed not to see Maria Clarkson.

"I just hope we aren't in here for too long," Billy said. "I'd hate to have to eat those rations for any length of time."

Paul Kashiyama, a large, muscular man with a crew-cut, spoke up. "Those rations are no worse than Squires' cooking."

Jamie wanted to ask him which bad movie he'd stolen that line from, but said nothing.

*   *   *

Jamie met Maria his first day on the job, after he'd almost gotten into a fight. He remembered it all too well. Jamie had run out of the kitchen upon hearing the clatter of dishes and cutlery hitting the floor. A wall of flesh stopped him before he'd barely taken three steps into the mess hall.

Paul Kashiyama grabbed Jamie by his apron. "What the hell are you feeding us?"

"Cajun stew," Jamie replied meekly.

"It's burning my goddamn mouth! What the hell are you trying to do, kill us?"

Jamie tried to peer around Paul's massive bulk. The diners he could see had odd expressions on their faces. "It's supposed to be spicy."

"Spicy?" Paul tightened his grip. "This isn't spicy, it's goddamn nuclear. What the hell did you put in this?"

"Well, the recipe does call for hot sauce –"

"How much?"

"I put six tablespoons –"

"Your idiotic recipe calls for six tablespoons of hot sauce!"

Jamie shook his head. "The recipe calls for three, but I always double up because –"

"That's enough, Paul," a female voice interrupted. "He's new. Cut him some slack."

Paul released his grip. "You watch it," he said, jabbing a finger into Jamie's chest. "You're going to be the death of us."

The woman coughed. "Get yourself a drink, Paul. And clean up the mess you've made." She was in her early thirties, of medium build, with long curly light brown hair. She turned to Jamie. "Are you all right?"

"Yeah."

"I'm Dr Maria Clarkson, the base physician."

They shook hands.

"Jamie –"

"– Squires. Yeah, I know. Our new cook." Maria grimaced and swallowed. "Did you really put six tablespoons of hot sauce into that stew?"

"Well, yeah. At my last job everyone complained my food had no flavour. I've doubled up on spices and condiments ever since."

"Your last job was where?"

"Canadian Pacific. Earth to L5 shuttle."

"There's no spin gravity on those shuttles, right?"

"No."

Maria nodded. "That explains it. I guess those terracentric cookbooks don't tell you that food tastes different in zero-g. Weightlessness redistributes body fluids. People tend to feel congested in the head, so food seems to have less flavour."

"Oh…"

"Don't worry, new guys are entitled to one non-fatal mistake. And don't let Paul scare you. He's all bluster."

Jamie could see Paul cleaning up his mess.

"Welcome to Maryniak Base, Jamie."

"Thanks."

Maria covered her mouth and coughed again. "By the way, may I have another glass of water?"

* * *

Maryniak Base was a mining facility on the lunar farside owned by the Alamer-Daas Corporation. Headquartered in Montreal, ADC's properties included three other commercial Moon bases, half a dozen Earth orbiting stations, and an industrial unit aboard the L5 colony. Maryniak produced titanium and iron extracted from lunar ilmenite for export to the burgeoning Lagrangian point settlements.

The solar storm lasted eleven hours before the United Nations Space Development Agency gave the all-clear signal. Jamie returned to the kitchen to find things exactly as he had left them. Using the back of a knife, he scraped the diced onions into the trash, and dumped the liquid eggs. He then got some garbage bags from the cabinet and walked to the refrigerator, a second-hand unit purchased by ADC at a former rival's bankruptcy auction.

There was a knock on the door frame.

"Mind if I come in and pick up the TLD?" Billy asked.

"Go ahead."

Billy walked to the wall beside the refrigerator and pulled the thermo-luminescent dosimeter from its bracket. The TLD was a stubby fat tube, about the size of a fountain pen.

"It's a shame to waste all this food," Jamie said as he surveyed the refrigerator's contents.

"Yeah." Billy held up the TLD. "But until I've had a look at these, we don't know if the food in the shielded logistics module is compromised. If it is, we'll be eating those disgusting rations from the shelter until the company bothers to send up a shuttle."

Crenshaw broadcast a briefing on the status of the base at the end of the workday. Being one of the largest common areas, the mess hall was a natural gathering place. A large post-dinner crowd gathered to watch the monitors.

"*On behalf of the company, I want to commend everyone on the manner in which we handled this emergency.*" Crenshaw's image was dotted with dark spots, indicating pixel dropouts from the radiation damaged CCD elements in her office camera. "*The good news is that the impact on production will be minimal. The total dose in the shelters was less than 12 millisieverts, and the reading in the logistics module was also within limits.*"

Jamie let out a breath. The food supply was okay.

"*Now, the bad news. The proton degradation of the solar arrays was severe. Output from the power farm is down almost twenty percent. In order to maintain production levels and have adequate battery margin for lunar night, there will be unscheduled brownouts of non-essential systems over the next several weeks.*"

Jamie spotted Paul talking to Maria. She seemed to be grinning at something he said. Jamie frowned.

"*The other major loss is the greenhouse. All the plants will have to be destroyed. This will impact atmospheric regeneration, requiring increased duty cycles of the metox canisters for CO_2 scrubbing...*"

Jamie tried to push Paul and Maria out of his mind, shifting his thoughts to the loss of fresh fruits and vegetables. He would have to adjust the menu to meet the nutrition requirements while maintaining variety.

"*...other than that, we fared well. Some of the essential electronics we couldn't power-down suffered single event upsets, but the redundant systems kicked in as designed. We should be fully back on our feet when the supply shuttle comes through next month. In the meantime, we have a business to run.*"

Crenshaw's image faded to black.

The people in the mess hall began to disperse. Jamie managed to recruit two of them to help transfer supplies from the logistics module. He'd asked their names, but promptly forgot them, and they took off immediately after the job was done.

Jamie activated his tablet to plan next evening's dinner. Suggested menus, based on UNSDA food guidelines, were uplinked by the company nutritionist in Montreal. But on-site cooks had wide latitude in meal preparation to accommodate local preferences and nutritional needs. Jamie scanned the proposed choices: macaroni and cheese, quiche Lorraine, or fish and chips. He called up the nutrient specs for the shelter rations. They were short of the 150 microgram UNSDA RDA for iodine, so that would have to be made up.

It would be fish and chips tomorrow night.

*   *   *

The stethoscope felt cold against his chest.

"Breathe in," Maria ordered.

Jamie inhaled.

"And exhale."

Maria removed the stethoscope from her ears. "How do you feel overall? Sleeping well, eating okay?"

"Sure, same as always."

"Most people were due for a check-up in a couple of weeks, but because of the storm Crenshaw and I thought it would be wise to do it now." She took his arm and wrapped the blood pressure cuff around it. "How do you like Maryniak so far?"

Jamie normally hated medical check-ups, but being able to spend some time with Maria made this one more than tolerable. "I'm having some trouble fitting in. Everybody seems to be in a clique or circle, and I feel kinda left out."

Maria inflated the cuff, opened the valve, and slowly released the pressure. "I felt the same when I first got here. Maybe it's the corporate culture. ADC doesn't have the best reputation. You've only been here a month. Give it time. Look at the chart on the far wall, please."

Jamie stared ahead as Maria shone a light into his eye. "I guess you're right. Maybe it's me. I've had trouble fitting in all my life."

"What do you mean?"

"My parents divorced when I was small," Jamie said, "but neither of them wanted permanent custody of me. So I grew up getting shuffled around between them and various relatives and family friends. I guess that conditioned me not to settle down anywhere. Even for college, I ended up quitting and reapplying at three different schools before I finally got a degree."

Maria turned off the light. "What was your major?"

"Business." Jamie blinked. "I hated it. My classmates were arrogant snots who liked to hear the sound of their own voices, and the profs were eggheads who never left campus but still felt qualified to lecture us on how the 'real world' works."

"I see." Maria handed Jamie a cup of water. "Swallow when I tell you to."

She stood behind him and placed her fingers on his neck. "Take a sip now, please." As he swallowed, she felt his thyroid for tenderness.

Maria took the cup and disposed of it. She then went to a cabinet and got a syringe. "I need a blood sample. Please put your arm on the side rest."

Jamie felt a prick as the needle went in.

"How'd you get from business school to cooking?"

Jamie sighed. "I met someone in college, but she wanted to stay in town after graduation and I wanted to try something else in another city." He shook his head sadly. "Maybe it was the way I grew up that made me feel so compelled to move all the time, but it broke her heart. Took me two years to realize I'd made the biggest mistake of my life, but by then it was too late. Next thing I knew, she was married."

"I know how you feel," Maria said as she withdrew the needle. "But how does cooking come into this?"

"It just came to the point where I figured the only way to make the hurt seem worthwhile was to keep moving, to go as far away as I could. Can't get much further away than space, right? Every facility out here seemed to need engineers, doctors, and cooks. I'm not an engineer or a doctor, but I like to think I became a pretty good cook back at the college co-ops, so here I am. A man's place is in the kitchen, right?"

Maria labelled the blood sample. "Well, I think you're doing a great job, especially with the crappy food the company makes you work with." She smiled. "You're fine. I'll call if there's anything you need to know about in the blood work."

A few years ago, Jamie's list of jobs in the commercial space sector would have been shorter by a third. Companies believed the only purpose of food was physical nourishment, so having a dedicated cook was considered an extravagance. But food provided psychological as well as nutritional sustenance. Many corporations, including ADC, learned the hard way when productivity dropped by almost forty percent. Taking a lesson from the old terrestrial oil platforms, space companies began hiring full-time cooks, and worker morale improved immediately.

Such was the importance of Jamie's role in psychological support that Crenshaw granted him a power rationing waiver to use the oven. Tomorrow was the birthday of Fred Sabathier, the Shift 3 foreman, and coffee cloud cake was his favourite.

Jamie had just poured the batter into a tube pan and put it in the oven when he got a call from Maria.

"Jamie, do you have a minute?"

He glanced at the timer. "Sixty-five, actually. What's up?"

"I need to talk to you about something."

Jamie frowned. It had only been a few hours since his blood test. She wouldn't be calling him unless something was wrong.

* * *

"Thanks for coming."

Nervously, Jamie took a seat. "So, what's wrong with me?"

Maria laughed. "Nothing's wrong with you! I just needed your advice on something. This is confidential, of course."

"Of course," Jamie repeated, visibly relieved.

"I just examined the rover crew. They were out on a two-week helium-3 assay at Mare Marginis, but got back to base just before the solar storm."

Jamie nodded. He'd heard that ADC was studying the economic viability of Maryniak harvesting helium-3 isotopes from the lunar regolith to feed the new generation fusion reactors on Earth. "Are they all right?"

"They're all complaining about being...constipated."

"Really?" Jamie raised his eyebrows. "When did this start?"

"A couple of days into their expedition."

Jamie thought for a moment. "Well, they've been eating the same things as everyone else since they got back, and I stick religiously to the UNSDA guidelines. Anything I make has enough fibre, believe me, and the rover rations are also supposed to meet UNSDA standards."

"Maybe they weren't eating regularly," Maria suggested. "That and stress can be causes as well. I mean, the stupid rover broke down halfway through their mission."

"Maybe…" Jamie rubbed his chin. "Do you have the serial number for the rover rations?"

Maria consulted her tablet. "51800-8493227."

"Can I use your credentials to tie-in to the company logistics database?"

"Sure."

Jamie linked in. "That's odd. Give me that number again?"

Maria repeated it.

"That can't be right. It looks like the number you gave me is for an EVA ration. Let me do a search."

A few minutes later, Jamie put down his tablet, slowly shaking his head. "You're not going to believe this. I think they stocked the rover with the wrong rations."

"What?"

Jamie read the screen. "8493227 is a type of EVA ration. According to this, they're eaten by crews on ships without spin gravity before spacewalks. They're high in iron and sodium but low in fibre, so they won't have to take a dump when they're outside. The correct ration for the rover should have been 8493277. Somebody screwed up."

Maria rolled her eyes in disbelief. "All right, I'll let them know."

"I can do up a high fibre menu for the next few days. How do garbanzo pitas sound?"

"Yummy. While you're at it, make some of your blueberry oat bran muffins for the next rover expedition."

"I'll ask for a power waiver for those muffins." Jamie looked at his watch and stood. "Gotta go. Fred's cake needs attending."

"Thanks for your help."

Upon returning to the kitchen, Jamie immediately knew that something was wrong. It should have been filled with the smell of freshly baked cake. Instead, there was nothing.

He turned on the oven light. "Oh, no…" He opened the door. The cake was flat. "Dammit!" A brownout must have hit the kitchen while he was gone. Fred Sabathier's cake was ruined.

* * *

Jamie was in a bad mood.

Crenshaw had denied him another power rationing waiver to use the convection oven, so for the birthday party they had to make do with a prepared microwave pie. Fred seemed not to mind, but Paul had made endless jokes to Maria about Jamie not being able to "get it up". Jamie ground his teeth as he stirred the pot of pea and broccoli soup. Given ADC's miserly pay scales, Jamie thought it was a miracle there weren't more people like Paul at Maryniak. He briefly considered trying to slip something disgusting into Paul's serving.

Jamie could hear snippets of conversation from the doorway to the mess hall. He thought he heard someone say "explosive decompression". Instinctively, he glanced at the red ceiling light. It was off. He turned down the heat, covered the soup, and made his way outside.

Nobody was eating. In addition to those in the mess hall for their scheduled dinner slot, others had come in from the corridor and were standing. They were all watching the monitor, which was tuned to CNN Interplanetary.

"...*details continue to emerge on the accident that occurred at Banting Station just under an hour ago...*"

Banting Station was one of ADC's Earth orbiting research labs that used the microgravity environment to develop new pharmaceuticals. Jamie stepped closer to the monitor.

"...*explosive decompression of the laboratory module...*"

The mess hall lights suddenly flickered, and the screen went momentarily dark. When the image returned, it showed a gash along the end cone of the lab module. Around the edges of the opening, serrated aluminium was bent outwards like a twisted, metallic flower.

"...*emergency bulkheads engaged, sealing off the rest of the station but trapping four researchers in the lab, who are now presumed dead. The rupture occurred near a docking port on that module, but no vessel was attached at the time...*"

Jamie spotted Maria and Paul. She put a hand on his shoulder, whispered something into his ear, then turned and strode quickly out the door.

He took a step to follow her, when a sudden wave of dizziness hit him. Instinctively, he grabbed a chair.

"Are you all right?" someone asked.

Jamie nodded. His light-headedness disappeared as suddenly as it had come. But Maria was gone.

* * *

Jamie didn't see Maria again for almost a week following the Banting accident.

The cause of the tragedy was found within days. As a matter of procedure, the UNSDA investigation team reviewed the maintenance records of the station. They discovered that seven years earlier, an automated cargo vessel had collided with the laboratory module during a botched docking, cracking the end cone pressure bulkhead but not breaching it. ADC dispatched a repair team to patch the bulkhead, but they took a tragic shortcut. The crew spliced on the reinforcing section with only a single row of rivets instead of a double, compromising its long-term structural integrity. Seven years of thermal contraction and expansion from orbital sunrises and sunsets every forty-five minutes took their toll. The bulkhead simply blew out.

The accident itself was bad enough, but news of the company's complicity in the tragedy further eroded morale at Maryniak. Jamie noticed people were eating less, leaving more on their plates for him to clean up.

He finally saw Maria seven days after the accident. She was sitting at a table in the mess hall after the Shift 3 dinner slot, sipping a coffee, alone.

"May I join you?"

"Jamie!" She gestured at an empty chair. "Please."

"I haven't seen much of you lately," he said. "You've been eating in your quarters?"

She nodded, and sipped her coffee again.

Jamie thought for a moment. "I know just the thing to go with that." A moment later, he returned to the table, his right hand hidden behind his back.

"What's that?"

Jamie whipped out a raspberry muffin, and with a flourish placed it on a napkin in front of Maria.

She bit her lip.

Jamie sat down, concerned. "What's wrong?"

"Rick Chang was one of the guys killed on Banting Station last week."

"A friend?" Jamie asked.

"My ex-husband."

There was a moment of awkward silence. "I'm sorry," Jamie said at last.

"I hadn't as much as gotten an email from him in over five years. Didn't even know he was on Banting until the news reports came in."

The lights dimmed momentarily before flickering back on. Jamie waited for Maria to continue.

"We met at a summer job in the university medical biophysics department. On my birthday, he came to my desk with this huge raspberry muffin he'd bought at the coffee shop in the bookstore. He took out a napkin and put it down, much as you just did, except he'd stuck a candle in it."

"He sounds like a sweet guy," Jamie said.

"He was a jerk."

Jamie's eyes widened.

"He could be sweet, sometimes. But overall, he was a really selfish person. He'd do things for me, but he'd only go so far until it started encroaching on what he wanted, and then it stopped. One day, he came home and told me he'd accepted a job with Honeywell-Dettwiler in Darmstadt. He never even told me he'd been applying for other jobs! Rick just expected me to follow him, just like that. It was all about him. So he went to Germany, I did not, and that was that."

Maria looked up at Jamie. "It seems we have something in common, don't we?"

Jamie started to reach over the table towards her free hand—but stopped. He felt a runny dampness in his nose. A red blotch appeared on the table.

"What the –" Jamie put a hand to his nose.

"Are you all right?" Maria grabbed some napkins and handed them to Jamie.

He nodded.

"I've gotten complaints about nosebleeds lately. Maybe we should ask environmental control to increase the atmospheric humidity," Maria said.

"Dat wud be a gud idee-uh."

* * *

"So there's three golfers: a priest, a chef, and an engineer. They're at this course, but it's very frustrating because the people in front of them are really slow and won't let them play through. So back at the club house, they ask the owner who were these jerks. The owner says, 'Oh, please try to be tolerant of them. You see, they're firefighters who put out a blaze here at the clubhouse last year. Sadly, they damaged their eyes saving the building and they're now legally blind. So, in gratitude we let them play here for free.' The priest says, 'That's terrible! I'll go back to my church and pray for them.' The chef says, 'What a heroic bunch of guys! If they come to my restaurant, they can eat for free.' Finally, the engineer says, 'Why can't they play at night?'"

Billy started laughing hysterically.

"That's a good one," said Suhana Aziz, a mass driver technician.

Jamie was seated to her left. "I've heard that joke before, except it was a doctor instead of a chef."

"Well, I thought I'd make a slight variation in honour of present company." Billy gave Jamie a pat on the shoulder. "Listen, you probably don't hear this much, but I just want you to know that I think you're doing a great job."

Jamie was surprised. "That means a lot to me, Billy."

"I'm also glad the company didn't cheap out and got us a real cook," said Suhana.

"Thank you," Jamie said quietly. He eyed the unfinished plates of baked chicken around the table. "Although, I guess I didn't do so good today."

"Oh no, it's not that at all!" Billy said. "I just haven't been very hungry lately."

The conversation was interrupted by a retching sound. All three turned simultaneously in the direction of the noise.

Sarah Schubert, the rover driver, was throwing up.

"It wasn't that bad," Paul called out as the pungent odour of vomit filled the room.

Billy and Suhana helped Jamie clean up the mess, amid a string of apologies from Sarah. She had no idea what had happened, but told them she'd had a queasy stomach for days. She promised to make an appointment to see Maria at the earliest opportunity.

Jamie went to the sink to wring out the mop. He was rolling up his sleeves, when he suddenly stopped and brought his arms to eye level.

"What the hell?"

The insides of his forearms were dotted with blisters.

*　*　*

"Do they itch?"

Jamie shook his head.

"Okay, if you don't mind, I need another blood sample."

As Jamie pressed the cotton against his arm, Maria gave him a small tube. "This cortisone should ease the blistering."

"Do you know what caused them?"

"No," she said, securing the cotton with a bandage. "How do you feel overall? Anything unusual you've noticed?"

Jamie thought for a moment. "Sometimes nosebleeds, like when I was with you in the mess hall a few nights ago. I've also been getting these sudden dizzy spells. Just for a moment, then it goes away."

"What about your appetite?"

"I don't think I've been eating as much as I usually do." He paused. "Actually, I've noticed people seem to be eating less in general."

"People all over the base have been coming to me with similar symptoms. Billy's got blisters like you. Others have been complaining about nosebleeds, loss of appetite, nausea, and vomiting. Even I've felt lightheaded sometimes."

"What would cause these symptoms?" Jamie asked.

Maria exhaled slowly, staring at the ceiling for several seconds before replying. "I probably shouldn't be telling you this, but…the symptoms appear to be consistent with low-level radiation sickness."

A knot formed in Jamie's stomach. "Who else knows?"

"Crenshaw, Montreal…and now, you."

Jamie's voice was trembling. "Wh-what do we do? Do we…are we all going to get cancer or something?"

"Let's not jump to conclusions. According to the monitors, we weren't exposed to a dangerous dose."

"What if they're wrong?" Jamie exclaimed. "After what happened on Banting, how do we know…what if the company didn't build the shelters to spec?"

"Both Sarah and I show normal white blood cell counts. That's the weird part. If it's radiation sickness, particularly sickness advanced enough that we're seeing vomiting and nausea, we should also have reduced white blood cells, but we both have normal counts."

Jamie calmed down, a little. "So what else could it be? Food poisoning?"

"You tell me."

Jamie thought for a moment. "It's not likely. The refrigeration systems in both the kitchen and the logistics module are fine. Almost everything I make is well cooked, especially since we lost the greenhouse. Also, we eat a wide variety of foods, so it can't be any one item."

"That's what I thought."

An uneasy silence fell between them.

"If it is radiation sickness, or even it isn't, you've got to tell people," Jamie said at last.

"We can't say anything until we know for sure," Maria said. "We'd cause a panic."

* * *

Jamie spotted the table with Billy Lu and Suhana Aziz.

"May I join you?"

Suhana looked up. "The chef graces us with his presence."

"Have a seat," Billy said.

Suhana poked her fork half-heartedly into her spaghetti.

"Something wrong?" Jamie asked.

"We almost had an accident in the field today," she said.

"What happened?"

"Freddie Wilson was out doing an induction coil change-out on the mass driver. The IVA guy, Grant McPherson, was supposed to have applied inhibits to the power bus before Freddie even went out, to give enough time for the capacitors to discharge. Except, he didn't. Caught his mistake at the last minute, thank goodness. I could hear him screaming on the loops, 'Don't touch the coil, Freddie! Don't touch the coil!" Suhana shook her head. "It was damn close."

"I know Grant," Billy said. "That's not like him at all. He's one of the most careful people I know."

"He said he was feeling tired, a little dizzy," Suhana said. "Just lost his concentration for a moment while going through the checklist."

Jamie looked at the unfinished plates of spaghetti. "How do you guys feel?"

"I don't seem to have much of an appetite. But your cooking's great, as usual," Billy added quickly.

"Sometimes, I feel like I want to throw up," Suhana said, "and I haven't been eating much either."

"Have you guys talked to –"

Jamie was interrupted by three short beeps, indicating the monitor was about to come on. Seconds later, Crenshaw's image appeared on the screen.

"This is a general announcement for all personnel. Staff are to report to Dr Clarkson immediately for medical evaluations. Individual appointments have been scheduled and will be pushed to your tablets within the hour. Every attempt has been made to accommodate shift requirements, but should you be unable to make your appointment please reschedule with Dr Clarkson at the earliest opportunity."

The mess hall erupted with noise even before the screen went dark.

"Something's wrong, and they're not telling us!" Jamie could hear Paul shouting over the commotion. "That solar storm did something to us!"

* * *

Jamie entered the infirmary, a tray of freshly baked cookies in one hand, a pot of coffee in the other.

"Chocolate chip?" Maria was impressed. "You got another power waiver for the oven?"

Jamie nodded. "Crew morale."

"Where the hell did you get the chocolate?"

"Base facility food manager's discretionary logistical supply," Jamie said as he put the coffee and cookies on her desk. "In other words, my own personal hoard. For special occasions."

"What's the occasion?" Maria asked before taking a bite.

"Our last week alive."

Maria almost choked on her cookie. "That's not funny!"

"Maybe I wasn't trying to be funny."

"Jamie, I don't know exactly what's going on yet, but I do know a few things. One thing is, we are not going to die...at least, not this week." She stared at Jamie. "Did you hear me?"

Jamie nodded slowly. "What else do you know?"

Maria grabbed another cookie from the tray. "I know you make great cookies, Mr. Squires."

* * *

Jamie did not sleep well. Over the past few days, he started having thoughts that somebody was tampering with the food. Twice he woke in a cold sweat, the second time going so far as to getting dressed and running out to check the kitchen. When he returned to bed, his dreams were of Paul...and Maria.

He woke up feeling nauseous. Much like a hangover, except that he hadn't been drinking. He wished that he had, because it would at least have made waking up like this worthwhile. The shower made him feel better, but his gums were tender when he brushed his teeth, and when he finished his toothbrush was pink.

Jamie stepped out of his quarters and headed for the mess hall. The corridor was practically empty at this hour. Such was the call of duty, to prepare breakfast for the Shift 1 crew.

He turned a corner—and was suddenly grabbed from behind and thrown against the bulkhead.

"Paul!"

The big man tightened his grip, pressing hard against Jamie's chest. "What's going on?"

"I don't know what you're talking about."

"You know damn well! People are sick all over base. Nobody's saying anything. But *you*..." Paul jabbed a finger into Jamie's chest. "I know you've been talking to Maria. We're all sick from the solar storm, right?"

Paul tightened his grip when Jamie did not answer. "Joe McKay barfed in his suit last shift. Pretty gross, huh? He's lucky we're on the Moon. If he'd been in free space he could've suffocated." He brought his face right up against Jamie's. "So, what is happening to us?"

"I don't know," Jamie said. "I'm sorry about Joe, but I really don't know. I haven't been feeling so hot myself. Why don't you ask Maria or Crenshaw?"

"Oh, I'll definitely be talking to Maria," Paul said. "But I thought I'd ask you first."

"Yeah, well I'm just the stupid cook, remember?" Jamie decided he'd had enough. "You've been on my case since I got here! You're just jealous because Maria doesn't –"

Paul raised his fist. "What the hell do you know about Maria, kitchen boy?"

"*Paul!*"

Jamie turned his eyes and saw Suhana Aziz.

"Leave him alone."

"He knows something!"

Suhana said calmly, "I know that I kicked your ass in aikido last year, and you can be damn sure I can do it again, right here, right now."

With a growl, Paul let go of Jamie. He glared at him for a moment, then abruptly turned and walked away.

"Are you all right?"

"Fine!" Jamie stormed down the corridor without thanking his rescuer. Marching right past the mess hall, he headed for the Beta sector habitation modules. He quickly found the room he was looking for, and pressed the door buzzer.

"Who is it?"

"Jamie!"

"Give me a minute."

Maria opened the door. It was clear he had woken her.

"No offense, but this had better be important." She looked him over, and her tone quickly changed. "Good grief Jamie, you're trembling. What happened?"

"You have to say something." Jamie's breathing was heavy. "You and Crenshaw, you guys have to say something."

"We can't make a public announcement until we know exactly what's going on. We'd cause a panic."

"There's a panic now!" Jamie snapped. "What the hell was Crenshaw thinking, making a public announcement for medical tests without saying why? People are scared. I'm scared, Paul's scared, we're all scared." He stared at Maria, his eyes pleading. "If there's anything for people to be scared of, at least let it be on the basis of facts, even partial facts, not rumours and hearsay."

Maria nodded slowly. "You're right. I'll talk to Crenshaw and Montreal about this."

* * *

It was an angry and frightened crowd that packed the mess hall to capacity. Those who couldn't make it in person were watching through the monitors.

Crenshaw had to shout to be heard, calling for quiet three times before she could speak. "You've all received the briefing material through your consoles and tablets, but I've called this meeting to personally answer any questions you might have on the current..." she hesitated "...situation, at Maryniak."

Jamie squeezed his way between people to put trays of sandwiches on the tables.

"Why won't the company come clean?" Predictably, the first to speak was Paul. "We're all sick from the solar storm!"

"That's not true," Crenshaw said. "The dose inside the shelters was within safe limits."

"What if the dosimeters were faulty?" asked Suhana.

"The TLDs are ancient technology, but they're reliable," Billy said. "I'd have preferred solid-state dosimeters, but the company likes TLDs because they're cheaper. In any case, I have no reason to think the readings are wrong."

"I don't believe anything you people are telling me!" Paul shouted. "How do we know the shelters were buried deep enough? How do we know there was enough shielding?"

"The storm shelters meet all applicable UNSDA standards," Crenshaw said.

"Do they? We all know how this sorry ass company screws up and cuts corners. Look at what happened on Banting. For God's sake, there's even a rumour they stocked the rover with the wrong rations!" Paul pointed at Jamie. "The company's too cheap to even hire a decent cook! I think they skimped on shelter construction, and Billy over there doctored the dosimeter data to cover it up."

"Are you calling me a liar?" Billy's face turned red. "Why would I go along with a cover-up? I'm sick too, you moron!"

"Everyone has a normal white blood cell count," Maria reported. "That is inconsistent with radiation sickness."

Paul didn't let up. "We have been exposed to a harmful dose! Everybody's sick. My hair's been coming out in clumps every time I shower."

Jamie fought the urge to make a sarcastic remark about Paul's hygiene.

"The TLDs don't lie," Billy reiterated.

"Then maybe kitchen boy's been putting something in our food!" Paul exclaimed.

Jamie decided to speak up. "Billy, are you sure the food in the logistics module wasn't compromised?"

"Yes," Billy replied. "The logistics module is shielded, just not to UNSDA human-rated standards. The food's fine."

"How do these TLDs work?" Jamie asked.

"They're tubes of lithium borate manganese." Billy held up his hand, with his thumb and index finger apart. "The crystals absorb energy from ionizing radiation. After exposure, I plug them in an analyser, where they're heated up to three hundred degrees Celsius. This causes the energy to be released from the crystals as photons. The analyser's calibrated to determine the total dose absorbed by the tube based on the light it gives off."

Jamie thought about Fred Sabathier's birthday cake. "Do you watch these tubes as they heat up?"

"Do I watch paint dry? Of course not. I usually step outside and do something else."

"What would happen if the power got interrupted as the tube from the logistics module was being heated up, before it got up to three hundred degrees?"

"Well, the TLD would cool down, and then when the power came back on they would..." A look of horror flashed across Billy's face. "Oh, crap..."

A silence fell over the mess hall.

"The food," Crenshaw said at last. "It's the food."

Paul said, "I knew it all along."

Nobody touched the sandwiches.

* * *

Maria sipped her coffee. "Quite a meeting, wasn't it?"

Jamie nodded.

"You helped solve the big mystery."

"All I did was ask a question."

"The right question," Maria said. "When the TLD from the logistics module was heated and cooled, it partially reset the crystals, so when it was heated up to its proper temperature the second time fewer photons were emitted, producing an erroneously low reading."

Jamie nodded again. "According to the nutritionist in Montreal, the radiation could've destroyed up to forty percent of the pyridoxine and thiamine content in our food." He shook his head. "We've been suffering from Vitamin B deficiency."

"I've prescribed mega doses of supplements," Maria said, "but everything in the infirmary got zapped worse than the food in the logistics module. The company's sending up a contingency supply shuttle, but until it gets here people are going to be popping pills like crazy to make up for the depleted dose in each capsule."

"I can tweak the menu," Jamie said. "Try to make the best of whatever Vitamin B is left in our nuked food. How does chicken and brown rice sound?"

"You're quite a hero, Jamie."

He laughed. "Will you stop that!"

"I mean it. You're a big part of the Maryniak team. I really hope you stick around." She glanced at her watch and drained the last of her coffee. "Gotta go."

"Hey, uh...How about dinner?"

"I'm always here for dinner," she said coyly.

Jamie grinned. "Would you like something other than chicken? I can make something else if you like."

"Surprise me." She smiled, touching his shoulder. "A man's place isn't just in the kitchen, you know."

Jamie watched her walk out of the mess hall. Smiling, he got up from the table and started towards the kitchen, all the while trying to decide what he would make.

Afterword

"A Man's Place" was first published in the anthology *Space Inc.* edited by Julie Czerneda, which was a collection of stories about future jobs in space. My original idea had been to write a story about an interior designer and the challenges they would face in developing habitats that were both functional and aesthetically pleasing in the challenging space environment. In the course of my research, I learned that some astronauts have found food to be less flavourful in space due to sinus congestion caused by the redistribution of bodily fluids in microgravity. Since my enthusiasm for space is closely rivalled by my love of food, I quickly changed the premise of the story to that of a cook on a lunar base.

Prior to the first human spaceflights in the 1960s, some doctors were concerned that people might not be able to eat at all due to speculation that reduced gravity could impair swallowing. Fortunately, the cosmonauts of the Vostok missions and the astronauts of the Mercury missions experienced no such difficulties because swallowing is enabled by peristaltic contractions in the oesophagus. The food provided to the first cosmonauts and astronauts was decidedly unappetizing, consisting of a semi-liquid mush that was squeezed out of toothpaste-type aluminium tubes. On April 12, 1961, cosmonaut Yuri Gagarin became not only the first human in space but also the first person to eat in space, squeezing tubes of beef and liver paste into his mouth. Astronaut John Glenn had applesauce on his mission in the same manner, while his successors on later Mercury missions ate compressed bite-sized cubes that were rehydrated by saliva as the food was chewed.

Today, cosmonauts and astronauts aboard the International Space Station (ISS) can choose hundreds of items from menus developed by their respective national space agencies, which can also be personalized with commercially available food products. As crews have become more international and diverse, the variety of food aboard the ISS has improved in ways that Gagarin and Glenn could only have dreamed about. Foods that have been available on the ISS include beef tongue with olives from Russia, yakisoba noodles from Japan, macarons from France, pemmican jerky from Canada, kimchi from South Korea, and saloona from the United Arab Emirates. Chinese taikonauts have on their missions enjoyed items such as yuxiang shredded pork, Kung Pao chicken, eight-treasure rice pudding, and herbal tea. Should the events of my story "Fixer Upper" come to pass, Chinese taikonauts might one day bring their food to the ISS as well.

If your mouth is watering at these descriptions, it is important to remember that good food in space is not a luxury but a necessity. Most cosmonauts and astronauts have experienced weight loss during their missions, and there have been reports of appetite being depressed in microgravity. If food is not palatable and tasty, then crews may not eat enough to get the required nutrition. Low zinc has been associated with exacerbating the reduced taste and smell from fluid redistribution in microgravity, which would affect overall dietary intake. Vitamin D deficiency is a concern because of the absence of natural ultraviolet light which normally stimulates this key regulatory factor in bone and calcium homeostasis. Suboptimal carbohydrate intake can reduce productivity and impede physical responses to emergency situations. On the other side of the food equation, excessive sodium intake during spaceflight is problematic because it exacerbates the microgravity induced calcium loss in bones, elevating the risk of kidney stones.

As space missions get longer, culminating in permanent settlements like the fictional Maryniak lunar base in the story, the role of food in providing psychological as well as nutritional sustenance will become even more important. On Earth, food is an integral part of social occasions like holidays and the commemoration of special events like birthdays and weddings, and the term "comfort food" can sometimes be quite literal in describing meals that provide nostalgic or sentimental value. These roles will surely be amplified during the long-term confinement required for extended space missions. Like a family having a conversation around a dinner table, cosmonauts and astronauts are encouraged to have regular meals together for team building and morale.

For the ISS, food is prepared on the ground and then launched to the station aboard cargo vehicles like the Progress, the HTV-X, and the Cargo Dragon. The ISS does not have refrigeration or freezer capabilities for food, so items need to be freeze-dried, thermostabilised, or irradiated to enable long-duration storage at ambient temperature. A limited quantity of fresh fruits and vegetables arrive with each cargo vehicle, and these are always sought-after and eagerly consumed by the crew. There is a story about American astronaut Shannon Lucid who, during her time on the Russian Mir space station, took a bite out of a raw onion that had arrived on a Progress cargo vehicle and declared it to be the best thing she had ever eaten.

The current system of depending on resupply from Earth for all food items is undesirable for a lunar base and would not work at all for longer duration missions such as those to Mars or the asteroid belt. These missions will require the creation of novel production systems that require minimal inputs while maximizing the output of safe, nutritious, and palatable food. One potential solution is plant-based bio-regenerative life support systems that would have the triple benefit of not only providing fresh food but also atmospheric regeneration and positive psychological support for the crew. The first Earth orbiting space station, the Russian Salyut 1 in 1971, had a small experimental greenhouse (appropriately named Oasis) in which the cosmonauts attempted to grow flax plants. A more advanced greenhouse called Lada (named for the Slavic goddess of fertility) has been operating in the Russian Zvezda service module of the ISS since 2002, supporting ongoing studies of plant growth in microgravity and the feasibility of producing edible vegetables. Of particular interest are space-grown cruciferous vegetables as a potential source of dietary phytochemicals for cancer prevention.

As humanity moves further into space, I hope that our culinary horizons will expand as well. Perhaps one day our descendants may enjoy such culinary delicacies as plomeek soup, stewed gagh, mapa bread, or iced raktajino. Bon appétit!

Plot Device

To: *"Mazotta, Kathryn M." <SingleID587932354>*
From: *"TL&L Editor" <AIReg24994701>*
Subject: *Acceptance of Story*

Thank you for submitting your work to Tales of Love and Loving *("the Zine"). We would like to publish your short story entitled "All Elaine Wants is Love" ("the work"). Click the hyperlink to download the contract. If you have any questions, do not hesitate to contact the Editor.*

Kathy Mazotta had finally done it.

Two years ago, she had lost her job at the news division of Paramount-Bell WorldMedia, another victim of the CAN tools that had swept the industry. There were virtually no jobs left in journalism that still required human writers. Kathy had been trying to break into the fiction market since losing her job.

Now, at long last, she had her first sale.

Kathy glanced at the clock. Sean would be home in a couple of hours, but this couldn't wait. It would be so ironic to call him at work.

* * *

"Hey, Kath! I'm home."

Sean Nolan walked into the study to find Kathy seated at the computer. He kissed her on the cheek. "Another story submission?"

"Uh, huh."

"Which zine this time?"

"*Tales of Love and Loving*. They just came online a few months ago." Kathy opened the mailbox. "Still nothing for you. You'd think those companies would have the courtesy to at least send an acknowledgement of having received your résumé."

He changed the subject. "Listen, how about a movie tonight? *Siege of Heaven*'s playing at the Dunbar."

"Sean, you know I don't like those violent –"

"Come on, it's just harmless fun." He pulled her closer. "Besides, isn't it the company that matters?"

They went to see *Siege of Heaven*. The plot was very simple. A group of terrorists had taken over Space Station Beta, so a commando team was sent to rescue the hostages. To complicate matters, the girlfriend of the team commander was also aboard Beta, and the man was plagued by guilt over the death of a previous lover in a similar crisis. But after a spectacular space battle that climaxed in a hand-to-hand combat sequence in which the commander killed the terrorist leader by smashing his head through a bulkhead (his head blew up in the vacuum, while the rest of his body still inside the module convulsed wildly) Beta was freed. The movie ended with a weightless love scene before fading to the credits, scrolled out over the strains of an adult-contemporary soundtrack song.

"What'd you think?" Sean asked.

"It sucked!" Kathy spat. "I'll bet a 10-year-old with a CAN program wrote that and made megabucks!"

"It's just a movie." Sean rather liked *Siege of Heaven*. The action was intense, the special effects were awesome, and the mixed cast of live actors and crofts was likable.

Each theatre exit was flanked by a smiling teenaged usher who recited the same words to every departing patron: "Thank you. Good night."

"Good night," Kathy responded.

Sean said nothing. Kathy gently elbowed him.

"Hey!"

"You're being rude."

"Oh, come on. They're *paid* to say that!"

"You're in Sales, Sean. You deal with the public. Don't you prefer dealing with polite people? I'd thought you'd be a little more sympathetic."

They made their way through the parking lot under a clear, cloudless night sky. Weaving their way through the cars, they eventually got to Sean's old Toyota.

"That really wasn't such a bad movie," Sean said.

"You have absolutely no taste, you know that?"

"Hey…" He gently pulled her face towards his. "Don't mock my taste."

Kathy abruptly pulled away. "How much longer do you think you're going to be at Lagoda?"

Sean sighed. "What kind of a question is that? Look, I'm sending out résumés, but with the economy the way it is…If people aren't hiring, they're not hiring. Look, Lagoda's not such a bad place – especially in Sales and Support."

"The faster you're out of there, the happier I'll be."

"Come on Kath, let's not start this again. I've been working at Lagoda since we met. You never complained until you lost your job –"

"Because of those damn CAN tools that *your* company puts out!"

"I have nothing to do with those products! Look, I'm sorry nobody's bought your stories yet, but if you want to work I've told you Lagoda's hiring writers to help refine the genetic algorithms that –"

"I don't want anything to do with them and I don't want *you* to have anything to do with them either!" She stared out the window into the night. "Not that I'd expect you to understand."

They drove home without another word.

*   *   *

Lagoda Technologies occupied a nondescript three-storey building at Main and National, off the eastern edge of False Creek, an oddly unglamorous location for the industry leader in computer-assisted narrative software. CAN was originally developed for use in high-end games, in which new story elements are generated in real-time to allow players a seemingly infinite space of plot trajectories. But as the technology matured, Lagoda began to spin-off CAN tools for real-world writing applications.

Sean's new cubicle was the same size as the one in his previous position, but at least this one had a window. He would have preferred one that faced False Creek instead of the rail yard, but it was better than nothing.

"How was your weekend?" asked Peggy Yang, one of the senior engineers.

"Not bad. Saw *Siege of Heaven*. It was pretty good."

"Yeah, my husband and I saw it last week." Peggy pointed at the picture on Sean's desk. "So, when are you going to get off your sorry ass and marry that poor girl?"

"Will you stop that?" Sean demanded as he sat. "You sound like her mother, always asking me when she's gonna get some grandchildren. Why do you think I hate going home?" He wagged a finger at Peggy. "And if you keep this up, I just might stop showing up here as well."

"Great! Better job security for me."

"Get outta here, you slacker. Some of us have work to do."

Peggy smiled. "Hey, speaking of slackers, did you hear about what happened to Vik?"

"Is he still complaining about not having a window? I told him the day he was hired: 'You're here so that *I* can leave!'" Sean laughed. "That old office was a dungeon. You have no idea how glad I am to be out of Sales."

"Real engineering work, better pay, window…Anyway, for once he wasn't complaining. Turns out absent-minded Vik left his briefcase on top of his car and drove off with it up there. The thing actually stayed put until he got onto the 99. Came flying off and hit a police cruiser, of all things!"

Sean was still chuckling as he logged onto his workstation. After spending a few minutes answering his messages, he opened a folder and began coding a dynamic library module for the company's special new project.

if (concept(i) = FALSE).and.(belief(character(i)))…

* * *

Kathy had locked herself in the study…again.

"How much longer are you gonna be?" Sean rattled the door knob. "We're supposed to be at your parents' place in an hour. You know how your mother freaks out if we're late!"

The door opened. Kathy glared at him before returning to the computer.

"How much longer?" Sean asked quietly.

"Never –" she raised an index finger solemnly "– *ever* interrupt a writer in the middle of inspiration. If I don't get this down, I'll lose it for sure."

"What are you working on?"

"I got a message from *Tales of Love and Loving*. They're interested in my story, but asked for some revisions."

"Wow!" Sean exclaimed. "Can I read what you've got?"

Kathy backed away to let him see the screen:

"Oh please, I love him so much!" Tears were streaming down Elaine's beautiful face. Her voice quivered with passion and sorrow. "Daddy, daddy…if you could only see, just how good he's been treating me…"

Tik-Gon, who was Elaine's father, was very angry. His face had a dark scowl. "No! I absolutely forbid you from seeing that white boy again! Over my dead body will you marry a gweilo!"

"No Daddy, no!" Elaine struggled to find a way to make her father understand. "Frank is so special. He's my man, the most beautiful man in the whole, wide world..."

Sean blinked a few times and swallowed before he spoke. "That's...good."

"It's all about empathy," Kathy explained. "A good writer has to get inside the characters, to understand their inner lives, to know what makes them tick." Kathy paused for a moment, then quoted, "'I have to have the character in mind through and through...I must penetrate into the last wrinkle of his soul.' Henrik Ibsen said that, and he's right. That's what those hacks and their fancy CAN tools will never get."

"Sure..." Sean said. "I'll wait in the den."

Kathy slammed the door.

Sean went into the other room, turned on the TV at a low volume, and started watching a sitcom. There was a running gag involving a whoopee cushion.

* * *

if belief(character(j)).and.(event(j)) then...

A tap on the shoulder startled Sean out of his concentration.

"Lunch?" Peggy Yang smiled. "Come on, don't make the rest of us look bad."

"I'm not hungry."

"Suit yourself." Peggy started towards the door. "Hey, isn't Hydro going to be doing that UPS generator test tonight?"

Sean looked up. "Oh my God, thanks for reminding me."

The last time B.C. Hydro did a test of the building's "uninterrupted" power supply, they somehow managed to take down the computer network, both the voice and vidphone systems, and disengage all the security maglocks. Sean made a mental note to remind Vik about his vidphone.

* * *

Kathy read the acceptance message again while waiting for the vidphone to connect.

"Hello," intoned the perfect image of a young woman.

"Uh...hello." Kathy frowned. She hadn't expected a croft to answer. Sean was almost always at his desk. She studied the computer-generated image for a moment. Yes, it was definitely a croft. She could tell from the eyes.

"I'm sorry, miss," it continued, "but I'm afraid Vikram Hakki can't speak with you at this moment. Would you like to leave a message?"

"What? Uh…sorry, I must have dialled wrong."

Kathy tried again, and was greeted by the croft once more.

"Hello, miss. I'm afraid Vikram Hakki cannot speak with you at the moment. Would you like to leave a message?"

"Vikram Hakki? Is this not the number of Sean Nolan?"

"It is not." The croft paused as the image refreshed. "Sean Nolan is no longer in Sales and Support. Do you want me to put you through to his new number?"

"No, tha –" She stopped herself from thanking the non-person. "Just tell me where Sean Nolan is now."

* * *

For years, Lagoda Technologies had been at the forefront of CAN software. Now, it was poised to take the next step. The company's goal was to develop a computer that could not only assist a writer, but would actually *create* its own tale. Lagoda would go beyond CAN into fully computer-*generated* narrative.

Sean surveyed the three linked workstations that formed a neural net known by the company codename of SHELLEY. The first machine, called PROUST, defined the initial theme, characters, and setting of a story. This would normally be done with a user input seed, but PROUST could also autonomously crawl the Web for story ideas.

The second computer, JOYCE, held a vast narrative parameters library that contained the complete vocabulary, syntax, and grammar of the English language. JOYCE also housed an extensive set of dynamic library modules. These DLMs codified such literary themes as love, revenge, jealousy, betrayal—and the one Sean was working on.

The last machine, KAFKA, was the heart of Lagoda's prototype CGN system. Using the structural data from JOYCE and the story seed from PROUST, it would run dozens of AI and CAN processes in parallel to produce the actual prose.

```
if (exist(action(i)).and.(!exist(state(q))).or.
(concept(i) = FALSE)
{
        if (belief(character(j))! = true)).or.(action(m))
                action(i) = action(m);
        else
                action(i) = NULL;
}
```

The message icon started blinking on the screen. He touched it.

"This is the front desk," a synthesized voice intoned from the speakers. "There is a person here to see you."

Sean frowned. He wasn't expecting any visitors. "Who is it?"

"The person identifies herself as Kathy Mazotta."

* * *

"How long?"

It was three in the afternoon, so the cafeteria was almost empty. Sean and Kathy faced each other across a table in a back corner.

"How long have you been working in R&D?" Kathy asked again.

"Seven months." Sean ground his teeth. It had to have been Vikram Hakki, forgetting to reinstall the message reroutes on his vidphone after the UPS test.

"You never had any intention of leaving Lagoda, did you? You weren't looking for another job, you just transferred –"

"What else was I supposed to do? There wasn't anything for me in Sales anymore! They've got crofts handling calls now. Hell, Vik Hakki's the only real person on shift during the day. The only other people are those who go out for service calls."

"And you couldn't do that?"

"I *hate* dealing with clients, Kath!" Sean shot back. "I'm an engineer, not a huckster. The only reason I stayed in Sales was to keep *you* happy. But R&D is where I've always wanted to be. It's where I belong, Kath! Why are you making Lagoda – making *me* – out to be such a villain? I'm sorry, but –"

"You're *not* sorry!" Kathy shouted. "You don't give a damn about how *I* feel. This is all about what *you* want again, just like always." She took a deep breath. "Ever read *1984*?"

Sean stared at her blankly.

"There's this character, this woman who works in the Fiction Department of the Ministry of Truth. She maintains these machines that churn out propaganda novels for the brain-dead masses." She jabbed her finger at Sean. "That's what *you* are. You and those computers, churning out junk –"

"*Computers* turning out junk? Didn't they have movies when you were little? I haven't noticed any change. If anything, I think stuff's gotten better!"

"Your damned technology's already shut writers out of non-fiction markets, and soon this CGN stuff will kill the fiction market too and then there won't be any decent stories –"

"Do you know how crazy you sound? Are you saying you'd rather I be out of work? Who the hell's paying the bills? I mean, it's not as if *you're* gainfully employed right now –"

"I'm out of work because of what *you're* doing!" Kathy exclaimed. "I tried to call to tell you I've sold a story. I've *sold a story*, Sean! But it took me two years to do it. Why is that? Why do you think I've only managed to sell just *one* piece of fiction in two years?"

"Well, maybe it's because *you're not a very good writer!*"

The instant the words left his mouth, Sean wanted to bite his tongue. "Oh, Kathy…what I meant was…"

Without a word, Kathy stood.

He reached out. "Kathy, I'm sor –"

She pushed him away. "Don't touch me!"

Sean watched her go, across the tables and seats, through the door, and out into the lobby, where she threw her visitor's badge at the security guard before exiting the front door. He watched her until he could not see her anymore.

* * *

if (plotelement(i)) = UNRESOLVED then prepare
(finalconflict)
        use (charactercrisis(random(1–10)))…

Sean's eyes glazed over the code. It had been almost a month, but he still couldn't concentrate on work.

"Hey."

He looked up. It was Peggy Yang.

"Look, if you want to talk –"

"Thanks." Sean's voice dropped to a near whisper. "I can't believe it's over. I mean, we always knew we were very different people, but I never thought…"

"I think you need some time. Why don't you take off for a while? You've got the flex hours."

Sean shook his head. "I think I'd rather be here. Keeping busy takes my mind off it a bit, you know?"

"Sure." Peggy nodded sympathetically. She gestured at his workstation. "By the way, what exactly is that DLM you've been working on?"

Sean stared at the screen for a moment.

"Deception," he said at last. "I've been working on deception."

* * *

Dear Dr Yaffe,

Thank you for sending me the latest story produced by Hemingway.3. *It is clear that your team has made significant progress since the V2.6 release.*

Given the limitations of the episodic experience database, it is not surprising that Hemingway.3 *again chose CGN as the topic of its story. What is interesting is how it was able to extrapolate a human conflict from the technological theme in a manner not seen in V2.6. This is likely the result of your implementation of the Bringsjord-Meteer heuristics, which also appears to have enhanced its capacity for characterization.*

It is also interesting how your redefinition of the narrative termination conditions have resulted in the tendency of Hemingway.3 *to produce ironic endings rather than the action-oriented conclusions of its predecessors. On a more serious note,* Hemingway.3 *still generates plots that have a certain algorithmic "connect the dots" feeling to them. Perhaps this could be remedied by decreasing the step size of the numerical narrative integration.*

Nevertheless, the story produced by Hemingway.3 *is a remarkable accomplishment. For the first time, a computer has produced a piece of fiction that appears comparable to one by a novice human writer. This is a significant breakthrough, and I look forward to the release of V4.0 with great anticipation.*

Professor Ian C. McCoy
Artificial Creativity Laboratory
Department of Computer Science
University of British Columbia

Afterword

In May 1997, an IBM supercomputer called Deep Blue beat the chess grandmaster Garry Kasparov in a much publicized match played in New York City. It was the first time a computer had defeated a world chess champion under tournament conditions. While some saw Deep Blue's win as a sign that artificial intelligence (AI) was catching up to the human brain, others played down the intellectual value of chess as a demonstration of true intelligence. From a purely logical perspective, chess is actually quite straightforward. A mindless computer could be made an invincible chess player as long as it follows an algorithm that traces out in a brute force manner all the consequences of each possible move until either a mate or draw position is found.

The nineteenth century mathematician Ada Byron, who upon her marriage became the Countess of Lovelace, suggested that creativity is the essential difference between the human mind and a machine. Byron argued that computing machines like the Analytical Engine she was working on with her contemporary Charles Babbage could only do what we program them to do but never produce anything entirely original because the latter requires the essential and uniquely human trait of creativity. A century later, the pioneering computer scientist Alan Turing devised his famous Turing Test (originally called the Imitation Game) in which a computer can be said to possess artificial intelligence if it could convincingly mimic human traits in a manner sufficient to fool a real person.

In 1980, the philosopher John Searle articulated the so-called Chinese room argument to illustrate what he perceived to be flaws in the Turing Test. In this hypothetical scenario, a person who does not speak Chinese sits in a closed room with a book of Chinese vocabulary and grammar. Another person who is fluent in Chinese passes notes written in Chinese into the room. Using the book, the person inside the room can select the appropriate response and pass it back to the Chinese speaker even though they do not understand Chinese themselves. The question posed by this scenario is whether the person in the room (i.e. the machine) actually understands Chinese or is merely simulating an ability to understand Chinese. Searle called the first situation "strong AI" and the latter "weak AI". Also known as artificial general intelligence, the concept of strong AI in which a machine is capable of human characteristics such as sentience, self-awareness, and even creativity has long been a staple of science fiction. Rather than chess or language translation, perhaps a more appropriate test of strong AI would be the ability to tell a story.

About a year after Deep Blue's defeat of Garry Kasparov, a team led by Selmer Bringsjord at Rensselaer Polytechnic Institute and David Ferrucci at the IBM Thomas J. Watson Research Centre announced the development of an artificial agent called Brutus.1 that could generate rudimentary short stories of less than 500 words.[1] The fictional Hemingway.3 in "Plot Device" was inspired by Bringsjord and Ferrucci's work on Brutus.1, which generated its stories based in part on mathematical codifications of literary themes such as love, revenge, jealousy, and deception. The comment in Dr Jaffe's closing letter that Hemingway.3 produces plots that have a certain algorithmic "connect the dots" feeling to them is a criticism that some have levelled at my stories!

Every year, thousands of aspiring (human) authors around the world take part in an event called "NaNoWriMo" (National Novel Writing Month) in which participants attempt to write a novel during the month of November. In recent years, a small community of computer scientists have engaged in a light-hearted sister competition called "NaNoGenMo" where the goal is to

---

[1] Bringsjord, Selmer. "Chess is Too Easy", *MIT Technology Review*, March/April 1998, pp. 23–28.

program a computer to write the novel for you.[2] Mark Riedl and his team at the Georgia Institute of Technology have developed a computer program called Scheherazade (named after the storyteller in the One Thousand and One Nights) that analyses crowd-sourced human accounts of everyday tasks and then attempts to produce simple vignettes about events like going to a restaurant or seeing a movie. But to be a true storyteller, a computer needs to "imagine" and not merely describe something. This is the goal of the What-If Machine (or the "Whim" as it is affectionately known), a project involving five European universities. Like Scheherazade, the Whim analyses databases of human prose and then twists what it has learned to produce an idea upon which a story might be based. At the University of Dublin, Tony Veale and his team have developed a program called Metaphor Magnet that produces nominally metaphorical insights by contrasting and inverting tropes harvested from social media.

Amit Gupta, who left a career as a Silicon Valley entrepreneur to be a science fiction writer, has developed an AI creative writing tool called Sudowrite.[3] Based on an autoregressive language model created by the OpenAI organisation called GPT-3 (Generative Pre-trained Transformer 3), Sudowrite generates human-like prose from whatever bits of text an author wishes to feed it to help "take your story in unexpected directions". As a test, I fed Sudowrite the opening sentences of "Plot Device" from which it produced ten samples of text that most certainly took the story in unexpected directions.

Three of the Sudowrite samples changed the nature of Kathy's writing ("It's not a story, it's a poem." / "Kathy's story is not a fiction story at all, but a news update describing the trial of Kathy's real-life nemesis, Mr. Dang." / "It is not a new novel but a book of remixes, essays on modern life made to look like fiction."), and one sample bizarrely changed the nature of her relationship with Sean ("Kathy and Sean had been having an open relationship, but Sean had been planning their wedding."). Two of the samples had implied elements of science fiction ("When Kathy called Sean at work to tell him what he'd just published, the caller ID display said that the call was coming in from the beginning of his life, long before Kathy had ever even met him." / "The man reveals himself as Sean himself inside a simulation."), while another two had disturbing elements of violence ("The news division announces Sean's death on-screen so that Sean's co-workers see him die in real time.") and even misogyny ("Sean is upset with the work and he

---

2 Meltzer, Tom. "Once Upon a Bot: Can We Teach Computers to Write Fiction?", *The Guardian*, 11 Nov 2014.

3 Marche, Stephen. "The Computers Are Getting Better at Writing", *The New Yorker*, 30 April 2021.

threatens to break Kathy's arms and legs and then hang her from the ceiling by her feet."). The remaining samples seemed to have almost random text ("Sean slams the phone down and tries to call the fire department." / "You're a deranged maniac! A liar, a thief, a liar, a thief, a liar, a liar, and a liar! A rabid mad dog!").

While machine-generated prose is certainly interesting, we are still a long way from artificial creativity. As a writer, I do not feel particularly threatened—at least, not yet.

Just Like Being There

Io was a bad place for something to go wrong.

The fifth of Jupiter's eighty moons, Io was deep inside a zone of radiation produced by solar particles trapped in the gas giant's magnetic field. Heated by the gravitational tug-of-war between Jupiter and its outer moons, the volcanoes that covered its surface erupted almost constantly, sending plumes of sulphur hundreds of kilometres into space.

Keith Mackay's spacecraft had landed on Io, and it was in trouble.

The spacecraft had touched down on a volcano called Emakong Patera with a not-so-soft landing that threw up a large cloud of sulphur. But according to the mission schedule, it should have landed ten minutes ago, well before Keith linked-in with the ansible. He immediately checked the seismometer, and one look at its peaks told him this was a bad time to visit.

"I have to get out of here."

"What's the problem?" crackled the voice of Flight Director Colleen Hodge through his headset.

"The volcano's gonna blow." Keith pulled the stick and the thrusters responded, lifting the spacecraft off the surface.

A split second later, the engines died. The spacecraft came crashing down, kicking up another cloud of sulphur.

"What's going on?" He pulled the stick again. Nothing happened.

"Keith, what's happening?" Colleen demanded.

"The engines aren't responding!"

At the edge of his field-of-view, Keith could see giant Jupiter hanging in the sky, its cyclonic Great Red Spot staring down at him like a bloody eye.

© The Author, under license to Springer Nature Switzerland AG 2002

E. Choi, *Just Like Being There*, Science and Fiction,

https://doi.org/10.1007/978-3-030-91605-3_10

The level on the seismometer continued to climb. He desperately tried to activate the thrusters again, but without effect.

Suddenly, the seismometer data spiked to a value that triggered a yellow alarm. The thrusters came to life, lifting the spacecraft away from the rumbling face of Emakong.

"Come on…"

The surface of Io was receding, but the numbers on the altimeter were climbing too slowly. Suddenly, all the alarms flashed red.

Everything went dark.

* * *

Keith ripped off his virtual-reality equipment and slumped in his seat. The alarms that flashed on the ansible only told him the obvious—the instantaneous communication link with TeleProbe-42 was broken. He stared at his console, but the blank screen offered no further answers.

"Keith, please report to my office," said Colleen over his headset.

He packed up his veergear, shut down the console, and quietly made his way out of Mission Control. Most of the virtual-reality astronauts, or "veernauts", were busy operating other TeleProbes across the Solar System. The few who weren't turned to watch Keith leave, nodding silent sympathy.

The walls of Colleen Hodge's office were covered with pictures of astronauts and a colourful assortment of mission logos. It was like a museum of the bygone era of human space exploration. At the corner of his eye Keith spotted the crew portrait for the Ares 7 mission, the one with his dad. He looked away.

"What happened?" Colleen asked.

"I don't know, but everything was all wrong from the start. According to the schedule, TP-42 should have landed ten minutes before, but when I established the ansible link it had just touched down. When I tried to take off —"

Colleen interrupted. "Why were you trying to take off so soon?"

"I told you, Emakong Patera was going to blow. I could tell from the seismometer data."

Colleen looked sceptical.

"I'm a geologist. I *know*." Keith continued. "So, that's when I tried to take off. But the thrusters wouldn't respond. I kept trying, but nothing happened until the yellow alarm went off. But by then, it was too late."

"Landing on a volcano, on the most volcanically active body in the Solar System." Colleen shook her head. "I knew it was a bad idea from the start."

"Emakong had been dormant for ten months. It's just bad luck it went active now."

"We can't fly missions based on luck."

"But I thought that was the point of using robots!" Keith protested. "We can take greater risks, go places we couldn't –"

"All right, all right." Colleen glanced at the clock. "We should be getting the radio data in another hour and forty-five minutes. We'll have a better idea of what happened then. In the meantime, I'm going to call in the failure investigation team. Unfortunately, this means you're off flight status pending the outcome of the inquiry. Don't take this personally, okay? Nobody's saying it's your fault, but –"

"It's standard procedure. I understand."

"In the meantime," Colleen continued, "I'm temporarily re-assigning you to the Public Affairs Office."

"Great," Keith muttered sarcastically. "Just great."

* * *

Keith's first assignment in his unwanted transfer to Public Affairs was to do a talk for students at Leslie Gelberger High School. While preparing for the presentation in his office, his phone rang. He glanced at the number and groaned.

"Hi, Dad."

Rob Mackay was 53-years-old, and had settled down to retirement in Canberra, Australia. Despite the years, he still resembled the stereotypical astronaut. His once brown crew-cut hair was now white, but he still had that steely-eyed, square-jawed look. Ten years ago, he had been the flight engineer on Ares 7, the last human Mars landing. During a solar storm, he had courageously stayed outside the ship's shelter to repair the engines in time for a crucial course correction manoeuvre.

"Hi, Keith." There was a split second time delay as the signal was relayed by communications satellite across the Pacific. "I heard about TP-42, and I wanted to see how you were doing."

Keith gnashed his teeth. NASA released news on the Web *way* too fast for his liking.

"I lost ansible contact while trying to abort the Io landing. We'll have a better idea what happened when we get some radio data back in about another hour."

"Well, these things happen," Rob said. "I remember when I was on Mars, and I accidentally –"

"You accidentally pointed the camera at the Sun and fried it. Dad, I've heard this story a thousand times." Keith frowned. "Are you saying I screwed up?"

"No! I was just –"

"Look Dad, I'm really busy. They stuck me in Public Affairs, and I've got to prepare this talk for some high school kids."

"I understand. I'm…I'm sorry to have bothered you." He smiled weakly. "Take care, son."

* * *

Twenty students attended Keith's talk at Gelberger High. Sixteen were physically present, while the remaining four were linked-in from remote sites. Bright, young, and ambitious, each of them came to hear about the dream of space exploration.

It was a dream that had almost died.

After close calls on the Ares 4 and 7 Mars missions, followed by the Clavius Moonbase disaster, the U.S. Congress had passed legislation prohibiting NASA from sending astronauts beyond Earth orbit. Out of the ashes of the human deep space exploration program, the TeleProbe project was born.

Keith began his talk. "The biggest problem in operating spacecraft by remote control is the time delay. A radio signal takes 2.7 seconds to go from the Earth to the Moon and back. For Mars, the round-trip delay can be up to forty minutes, and for Jupiter and the outer planets you're talking hours.

"To get around this, TeleProbes use two special technologies: the remote agent, and a machine called an ansible." Keith showed a flowchart. "The remote agent is a sophisticated computer program that allows a TeleProbe to fly itself for long periods without contact from Mission Control. We only need to give the TeleProbe general instructions, like 'land on that asteroid', and the remote agent plans, schedules, and then executes everything needed to make it happen. It can even recognize and avoid hazards."

Keith displayed a diagram. "The second special technology is the ansible. This machine gives us limited periods of instantaneous communication with the TeleProbe. By removing the time delay, we can do things like field geology or other complex, hands-on tasks by remote control – things that used to require the physical presence of an astronaut."

"How much time does an ansible give you?" asked a student named Vikhram Ganesh.

"The ansible works on a theory called quantum entanglement," Keith explained. "The TeleProbes carry a supply of entangled photons, or light particles, that are 'linked' in a quantum mechanical way to a corresponding supply in the ansibles at Mission Control. Instantaneous communication is possible until the states of all the photons have been determined, until they've all been used up. Depending on the data resolution, that's usually between two to three hours."

One of the remote-linked students, Susan Fontaine, asked a question. "What scientific instruments do the TeleProbes carry?"

"The TeleProbes use a simple design based on commercially available parts to reduce cost. Generally, TeleProbes carry scientific instruments like stereoscopic cameras, spectrometers, seismometers, radars...a standard set of science instruments."

Susan continued. "I saw this program on DiscoverNet the other night. They were interviewing a veernaut named Wendy –"

"Wendy Chong." Keith narrowed his eyes. Wendy had quit the veernaut corps last year. "What was the interview about?"

"She said the TeleProbes were too limited to do useful geological work. She talked about the time she was operating TeleProbe-34 on Mars and she discovered this unusual rock, but because of the sensor limitations she couldn't figure out if the rock was an igneous type, an impact breccia type, or a sedimentary type."

"Wow." Keith was impressed. "I'm amazed you know so much about geology."

"Thanks. Anyway, the problem was that each identification of the type of rock would have supported a different theory about the geologic history of the site. She said an astronaut could have identified the rock in a few minutes, like on Ares 7 when Marv Shipley and – hey, wasn't your dad on that mission?"

The eyes of all the students suddenly lit up.

"Yeah, my dad was on Ares 7," Keith said at last. He changed the subject. "Remember though, each TeleProbe costs ten times less than a human mission, so even if there are some limitations I'd say it's a good deal." He took a deep breath. "Believe me, when you're in ansible mode at the highest data resolution, with the full virtual reality gear, it's just like being there."

* * *

Upon his return to NASA, Keith went to Colleen's office to get the latest on the investigation. The news was not good.

"We've lost radio contact with TeleProbe-42," she said. "Taking into account the time delay, the signal ended at the same time you lost the ansible link. Also, the Shapley Observatory has confirmed that Emakong Patera did erupt."

Keith's chest fell. "But we have data up until then?"

"Yes." Plots appeared on Colleen's computer. "The engineers have started analysing this. So far, they haven't found any obvious mechanical malfunctions."

"Looks like some of the electronics were already starting to fail because of the radiation," Keith observed. "But the remote agent switched to the backups like it's supposed to."

"There's a chance TP-42 survived," Colleen said. "We're going to use the Deep Space Network antennas to try to re-establish radio contact over the weekend. But with forty-seven operational TeleProbes across the Solar System taking priority, we'll only be able to make periodic attempts."

"Is there anything I can do?"

Colleen shrugged. "Cross your fingers and hope for better news on Monday."

* * *

Keith couldn't wait that long. Throughout the weekend, he contacted NASA every couple of hours, trying to get the latest news. Finally, he got a call from Colleen ordering him to leave the engineers alone.

On Monday, as Keith was driving into the NASA parking lot, his phone rang. Recognizing his parents' number, he muted the handset. He'd get back to them when he had time.

Moments later in Colleen's office, he learned that crossing his fingers hadn't helped.

"The Deep Space Network wasn't able to re-establish radio contact," she told him. "NASA Headquarters has officially declared the TeleProbe-42 mission over. If it wasn't destroyed in the eruption, the radiation must have fried its electronics by now."

Keith couldn't hide his disappointment. "But we still don't know why I couldn't take off."

"No, but the engineers did find something odd." Coloured lines traced horizontally across Colleen's computer screen, like the vital signs of a patient. She pointed. "See that? An unscheduled thruster firing."

"That's too big for a course correction manoeuvre." Keith thought for a moment. "TeleProbe remote agents are designed to automatically avoid hazards. TP-42's agent knew it had to avoid flying through volcanic plumes. That looks like a plume avoidance manoeuvre."

"But according to the pictures from the Shapley Observatory, there weren't any erupting volcanoes along the probe's descent trajectory."

"So why did TP-42 fire its engines?"

"We don't know."

When Keith returned to his office, he found a message on his system from his mother. It was text only, just a few lines long, but he took several minutes to digest its contents. When he finally dialled the phone, his hands were almost trembling.

"Wooden Hospital."

"Could you put me through to Victor Poulos' room?" The pseudonym was no doubt intended to keep the reporters away, at least for a little while.

A moment later his father came on.

"Keith!"

"I...I got the message from Mom," he stammered. "Dad...you've got cancer?"

Rob Mackay nodded.

Keith felt confused and afraid. "What...What kind of cancer?"

"Thyroid," his father replied. "I guess I couldn't dodge the bullet forever. According to the doctor – Dr Albacea's her name – it's probably related to the radiation I got during that solar storm on the Ares 7 mission."

"What's going to happen?"

"Dr Albacea says they're going to map my DNA so they can genetically tailor the treatment for me."

Keith shook his head. "It's...it's just not fair. I'm getting so upset about losing a TeleProbe, me sitting here every day on a comfy chair in an air conditioned Mission Control, while you...you had to risk –"

"Keith, listen to me. What I did – what all of us astronauts did – was important. If you ask me, we're still needed out there. I have no regrets about any of it. Just to have the chance to stand on Mars, to look up in an alien sky and see two moons...I wouldn't change a thing."

He looked away for a moment. "I have to go. Dr Albacea's here. Looks like it's time for more tests."

Keith nodded.

"Thanks so much for calling."

"Dad..."

"Keith, whatever happens...I want you to know that I'm very proud of you." He smiled weakly. "Talk to you soon."

Keith just sat there, staring at the blank screen. *What is it with me and Dad? Am I mad at him for being away so much when I was small? But Dad was always there for me when it counted.*

Or...could I actually be jealous of him, because he's actually been out there while I just shadow box in virtual reality? Keith was overcome with shame.

Am I really so petty? Just because I'll never have a chance to see two moons in an alien sky…

Keith blinked. Mars had two moons, Phobos and Deimos. At last count Jupiter had eighty. The four largest innermost moons, including Io, were called the Galilean satellites after their discoverer, the Italian astronomer Galileo. The four largest moons…

He set up a simulation on his computer and ran it. The Galilean moons waltzed about Jupiter in their Newtonian dance. He zoomed in on TeleProbe-42's course and changed the field-of-view to that the spacecraft would have seen on approach.

There was the answer.

"Thanks, Dad." Keith wiped his eyes with the back of his hand, grateful to be alone.

* * *

"The failure investigation team has concluded that your theory is the most likely explanation for the loss of TeleProbe-42."

Colleen was telling Keith the engineers' findings, but he really wasn't listening.

"Europa, the next moon out, briefly appeared on Io's horizon in the field-of-view of TP-42's horizon scanner. This fooled the remote agent into performing a thruster burn to avoid what it thought was the plume from an erupting volcano. The course change was small, but it was enough to delay the landing by ten minutes and forty seconds.

"When you cut in with the ansible, the remote agent was still finishing up the landing sequence of its program. So when you tried to take off, it interpreted the thruster firing as a malfunction and shut them down. It finally commanded a take-off itself when the sensor data met its danger criteria – a level that was higher than yours, which was based on human experience. But by then, it was too late. TP-42 was probably destroyed in the eruption of Emakong Patera.

"So, that's it. A simple computer error ended your mission." Colleen shrugged. "It's unfortunate remote agent commands override ansible commands. The TeleProbe designers felt that since the spacecraft would know best what its immediate environment was like…"

"Every machine needs an off button," Keith muttered, "and someone to push it". He looked at his father's picture. "Does this mean I'm back on flight status?"

"Of course." Colleen frowned. "You've been awfully quiet, Keith. Are you all right?"

"My dad's been diagnosed with thyroid cancer."

"What?" Colleen exclaimed. "Why wasn't I informed –"

"I'm sorry." Keith stood up quickly. "I've got to go."

* * *

The suborbital flight from Houston to Canberra took forty-five minutes. Stuck in traffic, the ride to Wooden Hospital took an hour. The irony of the drive taking longer than the trip across the Pacific was not lost to Keith as the car pulled into the hospital.

He met Dr Penny Albacea in the hall outside his father's room.

"We've tailored the monoclonal antibodies to his genetic profile," she explained. "These antibodies deliver the chemotherapy directly to the cancer cells. So far, your father is responding well to the treatment."

"But if this doesn't work, his thyroid will have to be removed," Keith said. "He'll have to go on hormone replacement therapy for the rest of his life."

"That's the last option," Dr Albacea replied. "We'll do everything we can to make sure that doesn't happen."

"Thank you, doctor."

She gave Keith a pat on the shoulder before walking away. He watched her disappear down the corridor before opening the door to his father's room and entering.

"Keith!"

"Hi, Dad."

"Well, this is a pleasant surprise."

Keith was glad to see his father still had his hair. He pulled up a chair.

"How do you feel?"

"Well, I don't think the cancer will kill me," Rob said. "The food will do it first."

Keith smiled at the corny remark.

"So," his father said, "you're back on flight status."

"That's right."

"What's your next mission?"

"TeleProbe-56. It will arrive at Venus in February. Maybe I'll find out why the surface of Venus looks to be all about the same age. Some geologists think Venus sort of catastrophically erupts every couple of thousand years, resurfacing the whole planet with magma. Maybe I'll get a chance to see if this is true. Unfortunately, I'll only have half-an-hour to do it before the probe dies."

"You'll be lucky to get half-an-hour. Sulphuric acid clouds, hot enough to melt lead on the surface…Nice place to visit –"

"But I wouldn't want to go there!"

Father and son shared a moment of laughter.

"Dad…What was it like, to walk on Mars?"

Rob Mackay raised his eyebrows. "Oh, I'm sure you've heard me ramble about that a thousand times."

"No, I haven't," Keith said quietly.

He thought for a moment. "Well, I could move around very easily, since Mars' gravity is only about two-fifths that on Earth. It felt really comfortable, even in my heavy spacesuit…"

Suddenly Rob Mackay interrupted himself, and bowed his head for a moment. When he spoke again, it was in a very different voice. "I'll tell you what it was like. It felt like I was dreaming. I kept wondering when I was going to wake up. It was like…it was like when I was a little boy, and my Dad would tell me a wonderful bedtime story. Walking on Mars was the most wonderful story of my life."

Keith Mackay put his arms around his father, and held him tight.

It still took a person to be a hero.

Afterword

The rewards and risks of human versus robotic space exploration have been a topic of debate since the earliest space missions. In science fiction, the central role of human explorers is never in doubt, whether it's on the bridge of a starship or on the surface of a strange new world. But in the real-life exploration of space, the discussion is more nuanced.

Sending humans into space is expensive and risky. While funding varies from year-to-year, in general about half of NASA's annual budget is devoted to activities related to human spaceflight. To date, fourteen astronauts and four cosmonauts have lost their lives during space missions, and others have been killed during training and other activities. In "Just Like Being There", the U.S. government prohibits NASA from sending astronauts beyond Earth orbit following close calls on two human Mars missions (including the mission described in "Dedication") and a fatal disaster on the fictional Clavius Moonbase. Historically, the United States never seriously considered permanently ending its human spaceflight program even after the *Challenger* and *Columbia* space shuttle accidents, in part because of the political and socioeconomic considerations related to the high profile and prestige which these activities bring to the nation.

From a practical standpoint, astronauts have proven to be adept at deploying scientific instruments and repairing complex equipment in space. A well-known example is that of the Hubble Space Telescope, which was deployed from a space shuttle and was subsequently repaired and upgraded by astronauts over the course of five servicing missions. Following the *Columbia* accident, NASA considered either cancelling the final Hubble servicing mission or replacing it with a robotic mission, but questions about the feasibility of a robotic mission as well as pressure from both the scientific community and the general public convinced NASA to reinstate a crewed mission. Astronauts and cosmonauts have respectively repaired space stations on-orbit like Skylab and Salyut 7, and if my story "Fixer Upper" is to be believed a future expedition of Chinese taikonauts may attempt to revive the International Space Station.

Astronauts would also be expected to excel in field exploration on the surface of the Moon or Mars, taking advantage of human factors such as experience, judgement, and flexibility in the face of unknown or unexpected circumstances. The scientific return from the crewed Apollo missions was far greater than from the robotic Luna missions because the geologically trained Apollo astronauts were able to select the most representative samples from a site, recognize scientifically interesting geologic structures, and act upon their observations, in contrast to the Luna samples that were simply scooped up indiscriminately. However, as I described in the afterword to the story "From a Stone", the time the Apollo astronauts had on the lunar surface was so short (the longest cumulative extravehicular activity time was only 22 hours and 4 minutes on Apollo 17) and the mission schedules were so tightly planned there was actually little opportunity for spontaneous discovery. Perhaps this will change when humans finally reach the Martian surface.

No human has been on the Moon since Apollo 17 in 1972, and none have ventured beyond low Earth orbit since. So in order to explore the rest of the Solar System, our current choice is not between humans or robots but between robots or nothing. The exploration of Jupiter did not wait for the crewed *Discovery* of Arthur C. Clarke's *2001* but began with the robotic Pioneer 10 and 11 flyby missions in the early 1970s and continues today with the ongoing Juno orbiter mission. Since they do not require food, water or air, only a robotic spacecraft like New Horizons could have achieved the nine year, 4.7-billion kilometre journey to Pluto. Robots can also explore extreme environments that are far too harsh for humans such as the 465°C surface of Venus which even the Venera 13 lander could only endure for 127 minutes. Thanks to the robotic spacecraft of many nations, the initial reconnaissance of the Solar System has been completed.

One of the biggest challenges in the operation of distant robotic spacecraft is the time delay. The round-trip delay of a radio signal between Earth and Mars can be as long as 40 minutes, and the round-trip delay for the New Horizons spacecraft during its flyby of Pluto was nine hours. In "Just Like Being There", I postulate two technologies to get around this problem: a type of artificial intelligence called a remote agent, and a device called an ansible that enables limited periods of instantaneous communications. The word "ansible" was first coined by Ursula K. Le Guin in her 1966 debut novel *Rocannon's World*. To be clear, quantum entanglement cannot actually be used to enable faster-than-light communications because no new information is being transmitted between the entangled pairs, and any attempt to encode new information by forcing one particle into a particular quantum state would break the entanglement with the other particle.

The character of Rob Mackay was partially inspired by Gene Cernan, the commander of Apollo 17 and the last person on the Moon (so far). Mackay's closing monologue about walking on Mars feeling like a dream was based on a speech that Cernan once gave to secondary school students as described in Andrew Chaikin's book *A Man on the Moon*. While astronauts like Gene Cernan have been the most visible personifications of space exploration, it is important to remember that robotic missions are made possible by the dedication and efforts and hopes and dreams of hundreds of very human engineers, scientists, and technicians. Through streaming broadcasts and social media, we have watched their cheers and high-fives when the Perseverance rover landed on Mars, and we've watched their hugs and tears when the Cassini spacecraft plunged into the atmosphere of Saturn.

In the final analysis, all space exploration is human.

She Just Looks That Way

Thumbnail images of MRI brain scans covered the computer screen, a mosaic of Rorschach ink blots in greyscale. Each image represented a moment in time—a snapshot of a thought, a memory, a feeling.

Rick Park had given it much thought. He didn't want to change his memories, only what they meant. And that, he hoped, would change his feelings.

He turned to Dr Barbara Ho. "So, that's the only part that will be changed? This…um, fusion area?"

"The fusiform face area, yes." She selected one of the thumbnails and maximized it to fill the screen. "Located here, in the extrastriate cortex. It's where facial recognition and physical attractiveness are processed. We're been treating people with body dysmorphic disorder by modifying some of the neural pathways in this area." She moved the mouse pointer over the region. "People with BDD perceive themselves as ugly and disfigured, even though there's nothing wrong with them."

"But in your case…" She looked at Rick. "What you're asking for is…a little different."

"Can it be done?"

"It *can* be done, although we've never attempted it before." She paused. "Chris has warned you about the risks?"

"Yes."

"And you still want to go ahead with this?" She tapped a file folder on her desk. "We've got your signed consent and waiver, but I want to hear this from you myself. Do you understand the risks, and do you still wish to proceed?"

© The Author, under license to Springer Nature Switzerland AG 2014
E. Choi, *Just Like Being There*, Science and Fiction,
https://doi.org/10.1007/978-3-030-91605-3_11

"I understand the risks," Rick said slowly, "and I want to go ahead with the treatment."

"Fine." She opened the folder and started writing on one of the papers inside. "I'll say this much, you'll certainly be taking our research into a whole new area. Tell me, how long have you known Chris?"

"A long time, since the seventh grade," Rick said. "Didn't see much of him the last couple of years until he finished his masters at Wisconsin-Madison and moved here to Hopkins to work with you."

"Well, the next time you see him –"

"He's coming to pick me up after the treatment."

"Yeah well, when you see him, can you do me a favour?"

"What?"

"Tell him he still owes me a thesis." She closed the folder and put down her pen. "All right, let's get started."

Rick stood slowly, keeping his newfound doubts silently to himself.

* * *

The icon representing the LIDARSAT spacecraft traced a sinusoidal groundtrack across a Mercator projection of the Earth. In the bottom right corner of the screen, a clock raced ahead at many times normal speed.

Rick watched the orbital simulation with disinterest. His mind was elsewhere.

She would be here today.

Rick had known that she would be coming for weeks. He'd heard from his line manager, not her. When he found out he tried to email her, but she never responded. He called her up. Her father was home, said she was busy, couldn't come to the phone. She never called back.

The months apart without contact, or at least without any interaction initiated by her, only intensified his feelings. He told himself he was looking forward to seeing her again. He told himself things would be different.

"Rick?"

He turned—and there she was, at the entrance to his cubicle, standing beside the line manager, Harry Davidson.

"I'd like you to meet Mariel Beckenbauer, our new thermal engineer," Davidson said. He was in his late fifties, tall yet chunky, with a substantial beer belly his cheap suits couldn't hide. His grey hair was cut short, parted on one side, and he wore thick Coke bottle glasses.

"Mariel's from Canada," he continued. "She was a summer student here last year. It took a while for the TAA to go through the State Department, but here she is at last." He put a podgy hand on Mariel's shoulder. His fingers looked like dried sausages.

"Hello." This moment had been on his mind for weeks, yet now, it was all he could think of to say. He looked at Mariel, and a surge of emotions swept over him—confusion, longing, anger, regret, desire, sadness.

"Nice to see you," said Mariel with brittle neutrality.

"Come on," Davidson said. "I'll introduce you to the rest of the team."

As Rick turned back to his computer, a black depression settled on him.

* * *

The white saucer-shaped objected glided silently over the skies of Washington, DC. There were no visible markings or discernible methods of propulsion. From Rick's perspective, the object seemed to dwarf the Washington Monument and the trees of Gravelly Point.

He reached out to it.

That could be a you-pho.

Someone suddenly cut into his field-of-view, leaping upwards. An arm reached out and swatted the frisbee away, beyond Rick's grasp.

"All right! Way to go with the big D."

Another voice called from the side-line. "Force the line, Rick. No breaks!"

As the opposing player tapped the disc on the ground to put it back into play, Rick took up marking position. The opponent jerked his shoulder, apparently going for a forehand flick. Rick lunged to block, recognizing the fake a split second too late. The opponent went the other way, spinning around to deliver a backhand throw towards the end zone.

"Up! Broken!" Rick yelled, trying to warn his teammates he had let the handler throw to the undefended side of the field. But they were all out of position. He watched helplessly as the disc sailed to the end zone and was caught by a woman on the opposing team.

The game concluded, and the teams lined up on the field to shake hands.

"All right folks, good game," said Chris Brown, the captain of Rick's ultimate team.

Cassie Clarke glared at Rick. "We were counting on that force. You can't let them break you like that!"

Rick was in no mood to respond. Chris opened his mouth, but another teammate, Jill Kravitz, spoke first.

"Leave him alone, Cassie. We're just having fun here."

"Fun is fun, but I'd like to win sometime too!" Cassie retorted.

"Hey, hey." Chris raised his hands. "All will be solved with beer. Who's in?"

Cassie shook her head, and the rest of the team also declined.

"No beer for me, and besides my Dad is coming into town tonight," Jill said. She smiled at Rick. "But maybe we can hang out another time."

Chris watched the team depart. "The post-game drink is an essential part of the game. *This* is why we haven't won." He turned to Rick. "I guess it's just us, then?"

Rick nodded.

"God, you actually look like you need a drink. Where do you want to go?"

"Doesn't matter," Rick mumbled.

"How about Fadó? We can walk there from Gallery Place Metro."

The other team had already left the Gravelly Point field. Rick picked up his gear, and as he did so glanced towards a spot beside the Potomac River between two clusters of trees. A memory surfaced, and his heart tightened.

Fadó Irish Pub was surprisingly empty for a Friday night. Rick and Chris seated themselves at one of the small square wooden tables, ordering a pitcher of Yuengling and an appetizer of smoked salmon bites. In the background, "Why Don't You Love Me?" by Amanda Marshall played from the jukebox.

"So, this guy walked into the lab today," Chris said. "I swear he looked like a movie star. People must fall all over this guy, but he's got this thing about his nose and hands. Classic body dysmorphic behaviour. There's absolutely nothing wrong with his nose. Now his hands, sure, he's got dishwater hands from those two jobs –"

Chris tapped Rick's mug. "Earth to Rick?"

"What?"

"I wasn't going to say anything on the field, but Cassie was right. You *were* off your game today. Something wrong?"

Rick took his beer, knocked back a swig of it. "Mariel started working at Devcon a few days ago."

"Mariel?" Chris thought for a moment. "That girl from Canada you met last summer?"

Rick nodded.

"What happened?"

"I can't explain it. It's like she's a completely different person than the one I met last year. It's like a switch was thrown in her head – black is white, white is black, a hundred and eighty degrees. Her personality, the way she acts towards me, is completely different. I don't mean to be funny, but it's like she's been replaced by alien pod people, or she's a double from the *Star Trek* evil mirror universe or something."

"That's really weird," Chris said. "It must be hard to see her around all the time."

"Uh, huh." Rick wrapped his beer in both hands.

"How did you guys meet?"

* * *

The reception at the Goddard Visitor Centre following the Soffen Memorial Lecture was not well attended. Astronaut Shaun Christopher, the keynote speaker, had left immediately after his presentation, so most of the audience did the same. Surveying the stragglers, Rick spotted the usual suspects from the local aerospace contractors that served the NASA Goddard Space Flight Centre—Boeing, Lockheed Martin, Honeywell, Alamer-Daas, Raytheon. He was the only person from Devcon Systems.

Rick was cornered by some recent graduates of the NASA Academy who were fishing for job leads. He handed out his business card and made the obligatory polite encouragement to send him their résumés. Two of the alums made a show of scrutinizing his card, while the others politely thanked him and left.

That's when he saw her.

She wore a stunning blue silk dress cut in that pseudo-Chinese qípáo style that was popular of late, the kind with the long split up the side of the leg. Her curly brown hair flowed down to her shoulders, framing a cherubic face with dimpled cheeks. And she wore glasses. Rick loved women who wore glasses. The round wire frames perfectly accentuated her soft hazel eyes.

"Hello there!" she called out.

"Hello," Rick replied. Momentarily lost for words, he reacted with imbecilic instinct—he handed her one of his business cards.

She read the card. "Rick Park, systems engineer, Devcon Systems."

"That's me. And you are?"

"Mariel Beckenbauer."

"Mariel," Rick said. "Like Hemingway?"

"Like Cuba."

"Oh."

The NASA Academy grads introduced themselves and proceeded to ask her about job prospects. Rick took his eyes off Mariel and spotted some engineers he knew from the Applied Physics Laboratory. He excused himself. As he walked away, he heard Mariel explain she was also a summer intern.

Rick joined the APL group, who were discussing a reaction wheel problem on one of their spacecraft at Earth-Sun L1. It didn't take long for him to notice Mariel, again on the outside of the crowd. As before, she joined the group and introduced herself.

Rick said he needed a drink and excused himself, eventually joining another group. When she reappeared a third time, he finally figured it out.

"Cheers."

Rick and Mariel clinked their wine glasses.

"So, you're an intern?" Rick asked.

"Yes. I'm here for the summer."

"What are you working on?"

"I'm supposed to be doing thermal analyses for some Earth science missions over in Building 32, but so far I've only been doing PhotoShop stuff for Public Affairs."

"That sucks."

"I'm a foreign national," Mariel explained. "Thanks to ITAR, they won't let me do any real engineering work. My TAA's in limbo at the State Department."

"Where are you from?"

"Vancouver," Mariel replied. "But I'm living in Germantown for the summer."

By now, the wine bottles were empty, and the hors d'oeuvres trays had more toothpicks and used napkins than food. The reception was winding down.

"Can you give me a ride to the Metro?" Mariel asked.

"Sure."

It was a short ten minute drive to the Greenbelt station. In contrast to her manner at the reception, Mariel was strangely quiet during the ride. Rick tried to engage her in conversation, asking what she did in Vancouver, where else she had travelled in the world, what were her favourite movies. She didn't respond to any of his questions, and simply stared silently out the passenger-side window.

They arrived at Greenbelt Metro, and Rick pulled the car up to the Kiss and Ride drop-off.

"Well, that was a lovely evening with good company," Rick said.

Mariel nodded, and finally spoke. "You know, I haven't had a chance to see much of DC. If you have some time –"

"I'd love to."

* * *

The LIDARSAT preliminary design review meeting was a tedious affair that ended far past the scheduled time. When the presentations concluded, Rick made his way to where Mariel sat. He saw Davidson put a hand on her shoulder and say something. Mariel nodded, and Davidson left.

Rick was standing right in front of the seated Mariel. She was staring at her laptop, making no acknowledgement of his presence.

"Hi, Mariel," Rick said. There was no way she couldn't have heard.

Another person approached. Mariel look up.

"Oh, hello, Sanjay," she said. "I should have the new model runs for you tomorrow."

"No problem," Sanjay said. "See you later."

Rick watched the other man leave, then turned back to Mariel. She was on her computer again. He wanted to scream, to grab her by the shoulders and make her acknowledge him.

At last, Mariel closed her laptop and abruptly walked away without a word.

In a stupor, Rick followed her like an obedient puppy. She said hello to people she passed. Rick followed her upstairs and found himself outside her cubicle.

"Mariel, what have I done to upset you?"

* * *

The Washington Area Frisbee Club held an introductory clinic for novice players the first Saturday of each month at the Sligo Middle School. From the side-line, Rick and Chris watched a trio work the disc down the field in a weave drill.

"Put down that pivot foot, Kathy," Chris called out. "That's travelling."

"I finally had a chance to talk to Mariel," Rick said. He told him what happened.

"You asked her what *you* had done to upset *her*? Man, that's way more polite than I would've been."

"I suppose."

"And what was her explanation for all this? Wait, let me guess. Never really liked you? Met someone else? Thinks ultimate is for dogs?"

"No, none of that. But I wish it was, because it'd be a whole lot easier to understand."

"Then what the hell did she say?"

"She said...she said it's because she's gluten intolerant."

Chris turned to Rick. "She said that?"

"Uh huh."

"Okay...Those free doctors in Canada told her this?"

"She said she found out herself. She thinks she's been like that since she was born."

"Self-diagnosis. Gotta love the Internet. Actually, gluten intolerant people can become depressed, but you said she's only like that to you and normal to everyone else?"

"Yeah."

"I'd say she sounds bipolar, but who am I to argue with Google?" Chris shook his head. "Has she always been like this?"

* * *

"Where are we?" Mariel asked.

"Gravelly Point," Rick said. "Come on."

They got out of the car. He led her along a paved bike path that took them near the edge of the Potomac River. They found a spot between two clusters of trees. Rick produced a blanket from his knapsack and laid it on the ground.

"This is nice," Mariel said. "How do you know about this place?"

"My ultimate team plays here sometimes, on those fields behind us."

"Catching frisbees? Isn't that a game for dogs?"

"Yeah, except I don't use my mouth."

It was a clear night, and they had a perfect view across the Potomac. A Metrorail train was crossing the Long Bridge from DC to Virginia, a string of lights gliding over the river. Beyond the bridge were the illuminated Jefferson Memorial and Washington Monument. Far to the right was the dome of the Capitol Building, partially obscured by the darkened trees on the eastern bank of the Potomac.

It was Fourth of July in the United States capital.

Fireworks rocketed into the night sky to the left of the Washington Monument, exploding into showers of flame and colour and sparkle. Concussive booms echoed across the Potomac, and on the surface of the water, diffused reflections of the fireworks shimmered and danced in concert with the bursts above. In the background, the muted but audible strains of the *1812 Overture* could be heard from the radios of other spectators in the park and those on their boats in the river.

A plane roared overhead, taking off from Reagan National Airport. They looked up.

"That could be a ufo," Mariel said. She pronounced it like a word, "youpho".

"Well, there are some lucky people on that UFO with an amazing view."

They turned away from the sky, and towards each other.

Rick put his hand on her shoulder, ran it down her back. She didn't turn away. He pulled her closer. She put a hand on the back of his neck, pulling his head slowly towards hers. He moved with the touch, leaned in close. Their lips met.

The fireworks continued.

* * *

Chris and Rick were, once again, the only post-game patrons at Fadó Irish Pub.

"This is *unacceptable*." Chris pounded the varnished wooden table, jarring the pitcher and mugs. "Drinks are an essential part of ultimate. *This* is why we're losing."

Rick took a sip of beer. "Can you get me in for the treatment?"

"What treatment?"

"The treatment you and Dr Ho have going up at Hopkins, the one where you fix the brains of people who think they're ugly."

"You think you're ugly?"

"I *know* I'm ugly," Rick replied. "My mother said so once."

"Your mother – uh, never mind. Rick, that treatment is for people with body dysmorphic disorder. What's this got to –" Chris' expression changed. "This better not be about that girl."

"I'm sure I could get over Mariel if only…I didn't find her so damn attractive."

"What are you saying?"

"The treatment. You adjust the part of the brain that perceives attractive-ness –"

"The fusiform face area, yes."

"– so that these people no longer see themselves as being somehow disfigured."

"Right."

"So…" Rick continued, "It must be possible to turn it around. Let's say, you adjusted my brain so that…I won't find Mariel attractive anymore."

Chris didn't answer for a moment. "That's insane."

"Why?"

Chris grabbed Rick's mug, topping it up. "*This* is all the treatment you need."

"I'm serious!"

"The procedure is experimental, and we've never done what you're suggesting."

"I know it's experimental," Rick said. "I'm willing to be a guinea pig. Hell, I'm sure you and Ho could get a good paper out of this."

"Man, you are really screwed up." Rick opened his mouth, but Chris held up his hand. "Sorry, I'm not trying to diminish your feelings. But this is way overkill for a broken heart, and there are significant potential risks here."

"Like what?"

"Well, hypothetically…prosopagnosia."

"What's that?"

"Face blindness. The FFA is also the part of the brain that processes facial recognition. There's a chance you could lose your ability to recognize people."

Rick fell silent.

"Drink...your...beer..." Chris punctuated each word with a jab of his finger. "Women are supposed to outnumber men in DC, so do yourself a favour. Find another girlfriend!"

* * *

The rain was coming down hard. Rick had left his umbrella in the car. He grabbed a magazine, a month-old copy of *Aviation Week*, from a table before exiting the revolving door into the parking lot. Laptop bag in one hand and the magazine held over his head in the other, he made a dash for his car.

As he fumbled for his remote, he spotted two people leaving the Devcon building. It was Mariel, sharing an umbrella with Dan Ricardo, one of the division presidents. They jogged the short distance from the building to the reserved parking spot where Ricardo's silver Jaguar sat. The lights of the Jag flashed, and Rick watched them open the doors and get in.

The Jag didn't start right away. Rick saw them talking inside. Ricardo looked old. Mariel looked happy.

* * *

Rick and Chris sat at a table near the back of Clyde's Restaurant in Georgetown, under a skylight from which models of World War I airplanes were hung. Along one wall was a large fireplace, while another was covered with travel posters from the Twenties and Thirties.

"I took Mariel here once," Rick said.

Chris poked his appetizer with a fork, pretending not to hear.

Rick looked up. "I want it done."

"Want what done?"

"The treatment," Rick said. "I'm tired of seeing Mariel every day and being heartbroken. I can't take this anymore."

"The only treatment you need is to find a cuter dinner date than me."

"Chris –"

"Get over her!"

"It's not that simple."

"It *is* that simple."

"You don't –"

"Rick...grow up! I'm sick of hearing about this. Can we talk about *anything* else?"

"Some friend you are!"

"It's an experimental procedure with significant risks." Chris ticked off points on his fingers. "We've never done what you're suggesting. There's no way Barbara will go for this. Rick, we're treating people with serious neurological disorders, not – guys fixated on…wacky women from Canada."

"Spoken by you, who still has a thing for what's her name…Adrienne, from *undergrad*?"

"Hey –"

"You have no idea how much this hurts!" Rick's voice was rising. He looked around, and the other diners turned away, pretending not to hear. "To meet someone as beautiful as she is, to think you know this person…and then, suddenly, one day it's like a switch is thrown, and she becomes someone else. And every day, every time I see her, I hope that maybe that switch will go back and she'll be that sweet and wonderful person I fell in love with again."

Rick slumped in his chair. "The things I was going to do for her. Move to Canada, get a job in Vancouver…"

"Cook gluten-free food for her? Wake up beside her every morning and wonder if you've got Jekyll or Hyde for the day?"

They glared at each other. A waiter came and placed a pewter pitcher inscribed with the words "Refined Water" on the table. Neither man touched his food.

Finally, Chris spoke. "All right. I'll talk to Barbara tomorrow."

"Thank you."

"No, don't thank me, because it's not for you. It's for *me*." Chris wagged his fork. "I'm sick and tired of this girl and your moping. This'll be worth it if it'll make you shut up."

Rick said nothing.

"Besides, you're right about one thing." Chris poured himself a glass of water. "I'm sure we'll get a great paper out of this."

* * *

The treatment went faster than Rick had expected. When he woke from sedation, only five hours had passed. Dr Ho sent him away with a bottle of pills. "Retainers," she called them. He was to take one daily until finished, by which time the changes to the neural pathways in his brain would become permanent.

A beige Buick station wagon with wood-panel sides, a relic from the Eighties, pulled up to the curb in front of the Wood Basic Science Building. Rick got in.

"Hey."

"Yeah, hi."

Rick had barely closed the door and was still fumbling with the seatbelt when Chris put the car in motion. They turned left on to East Monument Street, then proceeded to the on-ramp for the Harbour Tunnel Throughway.

"How do you feel?" Chris asked.

"All right, I guess. But I have a bit of a headache."

Chris looked surprised. "Really?"

As they approached the Fort McHenry Tunnel, Chris slowed the car and rolled down his window to drop some coins into the toll bin.

"Take an aspirin when you get home," Chris suggested.

"Yeah, I think I'll do that."

They emerged from the tunnel and continued to Interstate 95. The dull buildings of Baltimore gradually gave way to the fields and farms of Howard County.

"Listen," Chris said, "I'm sorry about that time at Clyde's."

"Well, I can see why you'd get sick of hearing about my issues with Mariel."

"You went out to Vancouver last year to see her, didn't you?" Chris asked.

"Yeah. What a disaster that was."

"She did the Hyde thing up there too? Evil twin with the goatee from the *Star Trek* mirror universe?"

"You got it."

"Was she gluten intolerant then?"

"No. She told me she'd taken a bad fall while doing kung fu and hit her head."

Chris laughed. "I'm sorry!" He wiped his eye with a finger. "All I can say is, I would *love* to get this person into our functional MRI. Now *there's* a paper!"

Rick stared out the window at the passing countryside. "Speaking of papers, Dr Ho wanted me to bug you about your thesis."

"Consider me bugged."

Chris stepped on the gas, and the car surged down the interstate.

* * *

There was a knock at the entrance to Rick's cubicle. He looked up from his computer.

A young Indo-American man was there. Rick blinked.

"Hey, you gotta check this out," the man said. "Davidson and Mariel are having some kind of screaming match in his office!"

After a moment, Rick recognized the voice as Sanjay's.

He followed the man through a maze of cubicles. Davidson's office was floor-to-ceiling glass on two sides with vertical blinds that were currently open. Sanjay and Rick could easily see inside, so they dared not get too close as the opposite was also true.

Mariel and Davidson were indeed in the office. At Rick and Sanjay's distance they could hear little, but mouths were flapping and fingers were pointing. Rick could see some of the engineers closer to Davidson's office cautiously prairie-dogging over their cubicle partitions.

Suddenly, the door flew open and Mariel shouted, "Fine! I'll bring it up with Dan tonight!" She stormed out of the office and marched down the corridor.

Rick turned to the other man. He saw the other man, Sanjay, shrug his shoulders.

* * *

Time was running out.

The defender marking Rick was already at stall six. Rick had less than four seconds to pass the disc. He looked down field to the end zone. A blonde woman had broken away from her defender and was running for the corner.

"...stall seven, stall eight..."

Rick had a clear line of sight to her. She was looking right at him. But he hesitated.

"... stall *nine*..."

Rick was unsure, unable to throw.

"...*ten*. Disc is dead!"

Rick handed the disc to the opposing player, who tapped it into play. He had barely started counting stalls when his opponent threw a long hammer down field towards a teammate in the other end zone. The throw was completed, and the opposing team scored.

As Rick walked off the field, the blonde jogged up to him.

"What the hell was that?" she demanded. "I was going for the corner! Why didn't you pass to me?"

"Knock it off, Cassie," said another woman.

The woman named Cassie pointed at Rick. "He was looking right at me and did nothing! It's like he didn't know I was there."

"Well, maybe if you wore an orange shirt like the rest of us you might be easier to see," the other woman shot back.

"Why do you defend him, Jill? He's sucked all season."

"Hey!" A male voice chimed in. "Knock it off, both of you."

The woman named Cassie stormed off to get water. The other woman, Jill, looked at Rick with an apologetic expression.

Rick turned to the third speaker. After a moment, he said, "Thanks…Chris."

* * *

The Devcon cafeteria was serving tuna casserole. It was vile. A strange stench wafted from the plate. It was uneatable, so Rick didn't eat it.

He confronted the cook, a short middle-aged man with scrawny arms covered with tattoos who wore a small white apron and a silly, misshapen toque. His nametag read "Bobby Mac, Chef de Cuisine". Rick was convinced the cook was a parolee.

"I want a refund," Rick said.

"Why, sir?" Bobby Mac exclaimed brightly, a toothy grin plastered on his gaunt face.

"There's a weird smell. I can't eat it."

"But it's *tuna*, sir!" said the beaming Bobby Mac, as if it were an explanation.

After further negotiation, Rick got his money back. But he had lost his appetite. He looked at his watch and decided to return to his desk.

As he walked through the seating area of the cafeteria, he passed a table where a young woman was sitting alone.

"Rick!"

She got up and approached him.

"Hello," Rick said.

"Oh, Rick," she said, "I've been here for months but haven't spent any time with you."

"That's…all right."

"I haven't talked to you." She added, "I've been having many gluten attacks."

"You've been very busy, I'm sure," Rick said carefully.

"You need to find me. Yell out to me. Grab me by the shoulders." Her speech quickened with each sentence.

"I don't want to do that," Rick said.

"You are upset," she continued, her voice staccato, "because I am pretending with the others but being honest with you. It is easy to pretend. It's hard to be honest."

Rick had no idea what she was trying to say.

Suddenly, she threw her arms around him. Startled, Rick paused before returning the embrace. She held him for several long moments, gently running her fingers down his back.

Just for a moment, Rick felt a familiar frisson.

"I have to go," he finally said.

Rick let go of her hand, turned around, and headed for the exit. He walked slowly, as if waiting for someone to join him. When he got to the door, he was still alone. He put his badge against the card reader.

The door opened.

Rick walked through, and he did not look back.

* * *

The phone rang several times before Chris picked up.

"*Rick, what's up?*"

"There's something wrong with me."

"*So what else is new?*"

"I'm serious, man!" Rick gripped the phone tighter. "I've been having trouble recognizing people."

There was a pause. "*Say again?*"

"People at work, people on the team…it's like it takes me a second or two to see who it is. And then I bumped into Mariel at work yesterday."

"*And?*"

"And I felt…nothing," Rick said, a part of him knowing it wasn't quite true. "It was like she was a stranger."

"*What do you want me to do?*"

"Make an appointment for me to see Dr Ho again, right away!"

"*I'll try to get a hold of Barbara, but she's out of town until next week.*"

"Well, there's got to be someone else in the lab who might be able to –"

"*It would be best to wait for Barbara to get back.*" There was a pause. "*Listen, Rick…our game this afternoon, it's at the Reflecting Pool fields, right?*"

"Yeah, but –"

"*Meet me at Foggy Bottom Metro an hour before the game.*"

* * *

Chris and Rick rode the escalator out of the Foggy Bottom Metro station. As they emerged at street level, the entrance to the George Washington University Hospital appeared to their left.

"Think anybody in there can help me?" Rick asked.

"Sure," Chris replied. "There's lots of cuties at G.W."

"This isn't funny!" Rick snapped.

"Sorry," Chris said. "Look, I've spoken with Barbara. She can see you next week."

They walked south down 23rd Street, and as they passed the State Department the Lincoln Memorial, resembling a Greek temple with its limestone and marble faces and fluted Doric columns, came into view. They followed the throng of tourists making their way around the ring road to the front steps of the Memorial.

"Hey, the fields are that way." Rick pointed to the south side of the Reflecting Pool.

"We have time." Chris jerked his thumb at the Memorial. "Let's have a look."

"At Lincoln?" Rick asked. "Why?"

"Indulge me."

Puzzled, Rick followed Chris up the steps to the front portico. Dodging tourists, they walked past the massive Doric columns into the central hall, finding themselves before the sculpture of the seated Lincoln. The marble visage of the sixteenth president gazed unblinking towards the ivory needle of the Washington Monument in silent, benevolent contemplation.

Chris pointed. "The man was hideous, wasn't he?"

"*What?*" The chatter of tourists echoed loudly through the hall. Rick was sure he'd not heard right.

"Abe," Chris repeated. "He was one ugly dude."

"What are you talking about?"

"I read that one of his political opponents once accused him of being two-faced. You know what he said?"

"I have no idea."

"He said, 'If I were two-faced, do you think I would be wearing this one?'"

"What's you point?" Rick asked.

Chris turned back to the sculpture. "During his lifetime, Lincoln was widely regarded as ugly. But later on…well, what do you think of when you see Lincoln?"

Rick shrugged. "Emancipation Proclamation. Brought the country together after the Civil War. Is this a quiz?"

"Do you think he's ugly?"

"No, I don't think Abraham Lincoln is ugly!"

Chris nodded. "President Lincoln's physical features are beloved, not because of their physical qualities, but because of what they stand for."

"What's your point?" Rick asked.

"I worked on an interesting study when I was doing my masters at Wisconsin-Madison. We got the university's ultimate team together and asked them to rate each other on physical attractiveness. Then we got some strangers to rate the team members, but based only on photographs. You know what we found?"

Rick shook his head.

"There was a tight ass on the team, a Cassie Clarke type, that every other member of the team rated as ugly, even though I thought she was kind of cute myself. She's actually the reason I got into ultimate, but that's another story. Anyway, there was another woman, one of the team leaders, who was rated most beautiful by her peers. But the strangers rated the other woman more attractive on the basis of the pictures.

"Later on, we put some of the volunteers in a functional MRI, and it confirmed the results at the neurological level. The brain processes attractiveness differently when the person knows non-physical traits of the other person that are unknown or invisible to strangers. Non-physical factors are crucial to the subconscious assessment of beauty."

Rick waited for Chris to continue.

"So, it turns out the old saying was right all along. Beauty is a hell of a lot more than skin deep, and now we know there's an actual neurological basis behind it."

Chris looked his friend in the eye. "Rick, we never gave you the treatment."

It took a moment to sink in. "You, never –"

"Barbara put you under, you had a nice nap in our lab, and that's it. Nothing was done."

"But, the retainers –"

"I have no idea what those pills are," Chris said. "I hope they tasted good."

Rick was momentarily speechless. "You sneaky bastard."

"You're welcome. But still, I'll bet you don't find Mariel quite so hot anymore."

Rick thought for a moment. "No, I guess not."

"You know why?"

Rick said nothing.

"Because she's fucking insane."

Rick looked at his friend ruefully. "Is that a legitimate medical diagnosis?"

"Absolutely. Fucking insanity is a common affliction of many men and women, unfortunately. As for Mariel, I guess you could say…she ain't pretty, she just looks that way."

Rick put a hand on Chris' shoulder. "Thank you."

"What are friends for?" Chris looked at his watch. "Now, come on. We gotta win one for the Gipper."

Rick laughed out loud, to his total dismay. "Wrong president, dude!"

* * *

"Well everyone, we've got two big things to celebrate." Rick raised his beer. "Number one is, of course, our first win of the season!"

Shouts of "yeah!" resounded, and fists pumped the air.

"The second thing is…our peerless captain Chris has finally finished the first draft of his doctoral thesis!"

More claps and cheers. Someone yelled, "You the man, Chris!"

"Actually," Chris said, "there's a third thing, and that's our first full team post-game pub meeting. You have no idea how painful it was to be stuck with just *him* –" he pointed at Rick "– for company."

"*Cheers!*"

The team raised their beers again. Jill smiled, toasting her soda in Rick's direction.

Everyone had come out. The team had practically taken over the Froggy Bottom Pub's modest patio facing Pennsylvania Avenue. Even Cassie Clarke showed up. They had to pull together all the small square formica patio tables to seat everyone.

The revelry continued into the evening, but gradually people began to leave.

Rick checked his watch. "Well, it's been a blast, but it's getting close to my bedtime." He stood and waved. "See everyone next week."

As he walked by Chris, he patted him on the back. His friend smiled and nodded.

Rick had just stepped around the patio railing to get onto the sidewalk when a voice called out.

"Hey, are you going to the Metro?"

He turned and saw Jill Kravitz. "Yeah."

"May I come with you?"

"Sure."

It took less than five minutes to walk to the Foggy Bottom station, but it was still more time than Rick had ever really spent with her before. They had played together for months but had never really talked about anything except what was happening during a game.

Jill told him she was from London, England, and that she was working as an information researcher in the Geography and Map Division at the Library of Congress. Rick told her about his work on LIDARSAT at Goddard and the importance of satellite remote sensing to global environmental monitoring. They both agreed that *The Wrath of Khan* was without a doubt the best *Star Trek* movie ever. Rick resisted the urge to do the infamous Shatner scream.

They arrived at Foggy Bottom Metro and took the escalator down to the platform, where they prepared to part ways. She needed to take the Blue line to Franconia-Springfield, and he the Orange line in the opposite direction.

The lights along the edge of the platform began to blink, indicating the imminent arrival of a train.

"Hey Rick, are you doing anything this Sunday night?"

"No, not really."

"The Northern Pikes are at Wolf Trap," she said. "I've got tickets. Are you interested?"

"I'd love to. I haven't been to Wolf Trap in ages."

"Great! Call me."

"I will."

A puff of air hit them as the Orange line train emerged from the tunnel and rumbled into the station. Jill's dark hair flew up for a moment. They stood there, looking at each other as the train came to a halt.

Jill Kravitz smiled. She was beautiful.

Afterword

"She Just Looks That Way" was inspired by the work of anthropologist Kevin Kniffin, a former research fellow at the University of Wisconsin-Madison, and evolutionary biologist David Sloan Wilson at Binghamton University, who examined the influence of both physical and non-physical traits on perceptions of beauty. In their studies, volunteers were shown photographs and were then asked to identify which ones they found attractive. While physical traits such as a high degree of bilateral facial symmetry (for example, eyes that are identical in shape and size) have a strong initial impact, the results showed that non-physical traits have an equal or greater influence on perceptions of beauty over the long-term. In other words, people see attractiveness differently when they know other qualities of a person that are unknown to strangers.

To systematically consider the influence of non-physical traits on how people who are familiar with each other perceive physical attractiveness, Kniffin and Wilson conducted a series of studies involving both people who knew each other and those who did not. Volunteers were asked to rate physical attractiveness as well as non-physical traits such as talent and likeability, while strangers were only rated on physical attractiveness. For one study, the volunteers rated people based on photographs (including images of themselves) from secondary school yearbooks. In another study, members of a sports team rated images both of each other and of complete strangers. Finally, students in an archaeological field course were asked to rate each other on the first day of class and then six weeks later at the end of the programme. In each case, non-physical traits such as personality that would be known only to familiars had a more significant effect on the long-term perception of beauty than initial impressions based on simple physical attractiveness.

Kniffin and Wilson's studies provided illustrative cases of these findings. For example, a middle-aged subject who had not seen for decades a known person photographed in a secondary school yearbook responded with disgust when they recalled the person's character and described that person as ugly. In the sports team study, players considered a person with a poor attitude to be ugly and one of the leaders to be physically attractive, while strangers without knowledge of their personalities came to the opposite conclusion. After six weeks of working together on an archaeological dig, the perceptions of the students with respect to physical attractiveness changed based on their personal interactions over the course of the programme.

Other researchers have studied how attractiveness is processed by people with prosopagnosia or "face blindness", a condition in which a person loses their ability to recognize faces (in severe cases, even their own) due to brain damage resulting from a head trauma, an illness such as encephalitis, or insufficient oxygen supply at birth. A team led by Jason Barton at the University of British Columbia (UBC) studied individuals with prosopagnosia to gain insights on how the brain processes visual information and makes judgements about attractiveness. For those with prosopagnosia, the damage is usually found in the fusiform face area located in the extrastriate cortex. Since interpretations of beauty are largely based on culture-specific definitions of facial structure, it was thought attractiveness might be processed in this area. Although people with prosopagnosia cannot identify faces, they can still judge facial expressions such as raised eyebrows or pursed lips that express emotion and convey social cues. This observation led to a competing hypothesis that attractiveness is processed in the superior temporal sulcus region that reads changing facial expressions. Barton's team at UBC concluded that the same neural pathways in the fusiform region are actually used to process both identity and attractiveness.

Body dysmorphic disorder (BDD) is a condition in which a person perceives themselves as being ugly and disfigured even though there is nothing wrong with them. Individuals with BDD fixate on an imagined flaw in appearance or a slight physical abnormality, and often feel ashamed and depressed. To fix their "problem" those with BDD often pursue plastic surgery (sometimes repeatedly), and in serious cases are at increased risk of suicide.

About thirty percent of people with BDD suffer from eating disorders, which are also linked to a distorted self-image.

According to research conducted by Jamie Feusner, professor of psychiatry and bio-behavioural sciences at University of California Los Angeles (UCLA), the brains of people with BDD show no physical anomalies yet function abnormally when processing visual details. Using functional magnetic resonance imaging (fMRI) on volunteers while they viewed digital images of various faces through special goggles, Feusner's team discovered that people with BDD used their brain's left side (the analytic side attuned to complex detail) far more than those without BDD in the control group. Their findings suggest that BDD brains seem programmed to focus on facial details or even fill them in where they do not exist. Feusner team also discovered that the more severe the symptoms of a person with BDD, the more strongly their brain's left side activates during visual processing. These findings indicate that BDD has a physical basis in biology and cannot be solely attributed to societal attitudes towards appearance.

According to evolutionary theory, most animals (including humans) are attracted to those who are most likely to increase the likelihood of surviving and reproducing. In the case of humans, the fitness value of potential mates depends at least as much on non-physical traits such as intelligence and kindness as on physical factors. In a world where people are bombarded with harmful body messages from television and social media, these scientific results should encourage a wider discussion about what constitutes true attractiveness and perhaps even a reconsideration of practices such as cosmetic surgery.

Beauty really is more than skin deep.

Divisions

Château Laurier Hotel
Ottawa, Ontario, Canada
16:41, 12 June 1981

Sir John A. Macdonald, the first prime minister of Canada, stared accusingly at her as she tore the country apart.

This thought crossed Jennifer Simcoe's mind every time she looked at the painting. The trans-Canada railroad built under Macdonald by scores of forgotten Chinese labourers, the iron that united the nation, was just partitioned last month. She wondered what Macdonald would have thought of the cruel twists of history that had brought her—and Canada—to this place and time.

On 20 May 1980, the French speaking province of Québec voted to separate from Canada by a margin of 60.2%.

The official day of secession would be 24 June 1981—St. Jean Baptiste Day. In one year, all the legal and political mechanics of separation had to be worked out. The British North America Act had to be amended because it didn't have a provision for provincial secession. Other issues included the transfer of federal powers, an equitable division of federal assets and the national debt, and the formation of a new economic union.

Jennifer Simcoe was a seasoned negotiator, having secured agreements with the Soviets on wheat and India on CANDU nuclear reactors. Based on her record, Minister of State Bud Olson recommended Cabinet appoint her Canada's chief negotiator in the Division of National Assets negotiations, dubbed the "DNA Talks" by the media.

She got the job.

Jennifer took her eyes off Macdonald's portrait and focused them across the table. The Québec team seemed to be making a rather contrived show of scrutinizing the draft DNA Treaty. Her Québec counterpart, François Beaufrère, was shaking his head while his deputy Joël Campeau thumped his finger on a paragraph.

The one hundred and ninety-one page document was the culmination of seven intense months of negotiations. The final issue, the partition of Air Canada, was settled yesterday. All that was left was for the negotiators to sign two copies of the document, which would then be passed on to their respective Cabinets for ratification.

But something was wrong. They were focusing on one section. Jennifer strained her ears, but could not make out the whispered French between them.

"Hey, François?" A voice cut in. It was Jennifer's deputy negotiator, Dan McTavish. "Are we signing today or not?"

Joël looked up. "No."

Jennifer realized he wasn't joking.

"What are you talking about?" Dan snapped.

"We wish to make a change to this section," Joël pointed, "pertaining to the assets of Telesat."

"We settled Telesat last month!" Dan exclaimed. "I thought those issues were closed."

François Beaufrère smiled. It was the type of grin Jennifer associated with used car salesmen. She hated it.

"Yes, and we still agree with those terms," he said. "However, we wish to add –"

"The Anik Al satellite," Joël interrupted.

"*What?*" Dan shrieked.

François took a sip of water. "The current agreement deals adequately with Telesat's Earth-based assets. However, it is unfair that Québec be denied ownership of any of the Crown Corporation's space-based assets.

"To redress this imbalance, Canada must relinquish ownership of the Anik Al communications satellite to us."

Langevin Block, Parliament Buildings
Ottawa, Ontario
21:52

The now worthless pages of the draft DNA Treaty pertaining to the division of Telesat lay scattered across Jennifer's normally ordered desk. Under this agreement, Québec was to have gotten any Telesat land, buildings, teleports, and ground stations within its geographic boundaries. As for the satellites, the status quo would have been maintained until the launch of Anik C3 next year. At that time, six of the new satellite's Ku-band transponders would have been given to the new Commission de la Communication du Québec. Each transponder could carry four channels, and the CCQ—analogous to the Canadian Radio-television and Telecommunications Commission—would then distribute them to Québec-based users.

But now Québec wanted those terms changed. François offered to give up the six transponders on Anik C3 in exchange for possession of the entire Anik A1 satellite.

There was a knock at the door. Jennifer looked up from the papers. "Come in."

Dan McTavish entered her office. "I saw your light on. What are you still doing here?"

"Oh, I'm just looking over the old agreement, going over some research material, preparing notes on –"

"Why?" Dan asked. "That's Kevin Debus' job. He is our secretary, after all."

"Well, what are *you* doing here?"

"I'm single. I'm allowed to be out late." Dan smiled. "Look, there's not much more we can do tonight. I've contacted Telesat's counsel general."

"James Doherty? The guy who advised us on the first agreement?"

"Yeah. We'll meet him in his office tomorrow." Dan handed Jennifer her jacket. "Come on, it's late. I'll walk you to your car."

Kanata, Ontario
23:08

The porch lamp had been left on for her, but the rest of the house looked dark. At first, she thought everyone was asleep. But once inside, she saw the light on in the den. Entering the room, she found her husband Paul Tran watching the news.

An image of Québec provincial premier René Lévesque and some of his Cabinet ministers hosting an Algerian trade delegation flickered on the screen.

"Why didn't you call?" Paul asked.

"I was busy," she replied with barely concealed annoyance. Paul was an air traffic controller, a job that required him to work odd hours, too. Sometimes, he would be late coming home because, he claimed, he had been concentrating so hard on his shift he forgot where he parked his car.

"Look, you don't need to wait up for me. I've always told you that."

Paul sighed. "I left you some dinner, if you want it."

Jennifer nodded.

He left for the kitchen. Jennifer sat and watched a report on the continuing armed standoff between an Inuit band in Natashquana and the interim Québec militia.

Paul returned with dinner. Jennifer accepted the fried rice without a word and started eating. As she wolfed down the food, she thought she felt an insect tickling her neck. It took her a moment to realize it was her husband.

"Paul..." She shook him away. "I'm really tired, OK?"

He sighed and got up. "I'm going to check on Dill." Dill was their nickname for Dillon, their three-year-old son.

She turned her attention back to the TV. The latest paparazzi shots of Prince Charles and Lady Diana's wedding preparations were being shown. Jennifer watched this, and the rest of the newscast, alone.

Telesat Headquarters
Gloucester, Ontario
10:05, 13 June 1981

James Doherty ran his podgy hand through his short russet hair. "Let me get this straight. Québec's willing to give up the six transponders on Anik C3...to get A1?"

Jennifer nodded.

"Anik A1..." Lost in thought, James stared off at a child's drawing pinned up on his bulletin board. It was a crayon rendition of a spaceship, complete with tanks, fins, windows, and astronauts tethered with ropes.

"Well, what do you think?" Dan sounded impatient.

James messed his hair again. "This whole separation thing...Québec's made some pretty weird demands, but this one takes the cake."

"What do you mean?" Jennifer asked.

"Well, I don't understand why they'd want to do this. Anik A1 was launched in 1972. It's already past its design lifetime. It was only supposed to last seven years, but our engineers – miracle workers, those guys – they've squeezed some extra life out of it. But A1 won't last beyond next year."

"Is there anything special about that satellite?" Jennifer asked.

James frowned. "Not that I can think of."

Dan asked, "When will Anik C3 be launched?"

"Next year, on the Space Shuttle. Brand new bird. Ten-year lifetime, minimum."

"So, they want to give up capacity on a new satellite for an old one that's only good for a year. Sounds like a good deal." Dan turned to his colleague. "Jennifer?"

"I was just thinking about what my parents used to say about things that seem too good to be true."

Launch Complex 17-A
Cape Canaveral, Florida
12:49

Venting gases made the Delta rocket hiss and groan like a living creature. As tall as an eleven-storey building, with nine boosters clustered at its base, the expendable launch vehicle was an impressive sight. It was also doomed. In a few short years, its kind will be made obsolete by the Space Shuttle. The Shuttle will be the cheap way, the safe way, and the only way to orbit.

At 12:49:01, its RS-27 main engine ignited along with six of its nine Castor IV solid rocket motors to lift the vehicle off the pad. Umbilical lines fell away as the rocket rose on a column of fire and smoke, shaking the Earth goodbye.

The Delta executed a pitch manoeuvre and headed out over the Atlantic. Fifty-seven seconds later, the first six solids burned out and separated from the still climbing rocket. The remaining three were now supposed to ignite.

One did not.

The main engine gimballed its nozzle in a vain attempt to compensate for the malfunction. It wasn't enough. The Delta jerked off course, its payload shroud buckling under the aerodynamic load. Bits and pieces fell away into the slipstream as the rocket tumbled out of control. Through a wire running its length, the on-board computer sensed the break in continuity and detonated the range safety explosives.

Kanata, Ontario
11:53, 14 June 1981

"Bring the water to mommy, Dill," Paul instructed. "There's a good boy."

Jennifer looked up from her desk. "Oh, thank you." She took the glass out of his small hands and put it on her desk before picking him up. "So, whatcha gonna say now?" She pinched his cheek. "Whatcha say to mommy?"

Dillon squirmed. "I don't know."

Jennifer stroked his brush-cut brown hair. "Come on, Dill. What haven't you said to mommy today?"

He thought for a moment. "I love you, mommy."

Watching from the door, Paul chuckled. "I'll be downstairs. Lunch will be ready soon."

"Hey, wanna play with mommy? Bring Rufus. Mommy will play with Rufus and Dill, OK?" Rufus was a stuffed bear.

The phone rang. Jennifer gave Dill a hug before letting him down. The boy waddled out of the room.

"Hello?"

"*Jennifer Simcoe? It's Jim Doherty calling, from Telesat.*"

"Yes."

"*Look, I hate to bug you on a Sunday like this, but something important has come up. I'd prefer to talk about it in person. Could we meet later this afternoon?*"

"OK, I'll come right now."

A pause. "*Well... really, there's no rush. Actually, we just got home from church —*"

"Look, now's not a problem for me, OK? I'll come right to your office." She hung up before he could answer.

Jennifer gathered her material and stuffed it into her briefcase. She started for the door, but then she stopped.

Dillon was standing there. With Rufus.

Jennifer sighed. "Mommy has to go now, OK?" She gave her son a quick kiss on the forehead. As an afterthought, she gave Rufus a kiss, too.

Telesat Headquarters
Gloucester, Ontario
13:06

A security guard let Jennifer in and ushered her to a seat in the lobby. James arrived a few minutes later and informed her that Dan would be late. The pair proceeded to his office.

"Did you hear about the rocket accident yesterday?" James asked.

"No. There wasn't anything on the news."

James grumbled something. Aloud he said, "Anyway, this rocket was carrying the SBS-2 satellite, which was owned by an American company called Satellite Business Systems. This accident's put them in a real bind. All the capacity on SBS-1 is booked, and SBS-3 won't be ready until at least next year."

"What does this have to do with us?"

"Last night, representatives from SBS contacted us with an offer to lease any excess capacity we've got on our satellites."

Jennifer thought for a moment. "Excess capacity? That means –"

"Yeah," James nodded. "The only spare capacity we've got right now is on Anik Al."

"Well... this certainly adds a new wrinkle to the negotiations."

"I'll bet it does. But there's more." He handed her a sheet. "This is a list of the customers that had reserved transponders on SBS-2. Recognize anyone?"

Jennifer scanned the page, and her eyes widened.

Château Laurier Hotel
Ottawa, Ontario
13:56, 15 June 1981

"Why have you been negotiating with Satellite Business Systems behind our backs?" Jennifer demanded.

François spread his hands. "It was a simple business transaction between our government and a private company. We thought it did not concern you."

"It does concern us because it directly impacts these talks!" Dan snapped. "Why did you need to lease capacity on an American satellite?" He paused before answering his own question. "Because you know Anik A1's only got another year, and you need to make up for the transponders you're giving up on Anik C3. Am I close?"

The Québec committee's silence told him he was.

"There's a simple solution to this," Jennifer offered.

"What is it?" Joël asked.

"Go back to the old agreement. Maintain the status quo, and when Anik C3 is launched, you get six new transponders."

François nodded. "You're right. We'll go back to the original agreement."

The Canadian team's relief was short lived.

"Provided you also relinquish ownership of Anik Al to us," Joël added.

Dan slammed his fist on the table. "You guys want your gâteau and eat it too? No way!"

"Your terms are unacceptable." Jennifer narrowed her eyes. "Telesat stands to make a lot of money from SBS. Industry Minister Herb Gray is going to approve the deal tomorrow."

"We have a right to those profits. Québec helped finance that satellite," Joël noted. "We have a right to Anik Al."

"No," Dan said.

François said, "You won't reconsider?"

"This is it, François," Jennifer replied.

Joël passed a document across the table.

"What the hell is this?" Dan demanded.

François opened his mouth, but Joël spoke for him. "As you may know, the Anik D1 satellite is undergoing final assembly at the Spar Aerospace facility in Ste-Anne-de-Bellevue. This morning, Industry Minister Rodrigue Biron filed this injunction in Québec District Court barring the delivery of Anik D1 to Telesat."

"*What?*" Dan exclaimed.

François looked pained. "I'm sorry."

Langevin Block, Parliament Buildings
Ottawa, Ontario
18:11

Jennifer phoned James Doherty to brief him.

"*This is insane!*" James sounded hysterical. "*Anik D1 was just a week from being moved to the David Florida Laboratory for final pre-launch testing. This injunction's a low blow, Jennifer. We need to get that satellite to DFL now, or we won't make our launch date. NASA's booked solid. If we miss our flight, who knows when we'll fly? If we jump ship to Ariane, we'll be paying penalties through the nose. Regardless of who we go with, any delay will leave our customers without a satellite. We're stuck!*"

In seeking the restraining order, lawyers representing the Québec government argued that Anik D1's traveling wave tube amplifiers, built by MITEC in Pointe Claire, employed "sensitive technologies". Under new trade regulations imposed by the governing Parti Québécois, Québec-based companies exporting products classified as such are supposed to get the approval of the Québec industry minister.

"What about Anik C3?" Jennifer asked. "Is it safe?"

"*Yes. The C-series are built by Hughes in the U.S. But Spar Aerospace is the prime contractor for the D-series.*" James snickered. "*Buy Canadian, eh?*"

"Jim, it looks like they want their hands on Anik Al by hook or by crook. But why? Why are they so desperate for that old satellite?"

"*I'm... I'm not sure.*"

"Do you have any ideas?"

A pause. "*Yeah, maybe.*"

She gripped the phone tighter. "And?"

"*Look, I have an idea... but there are some people I need to contact first. I'll get back to you as soon as I can.*"

Langevin Block, Parliament Buildings
Ottawa, Ontario
20:47

After eating her dinner of a Mars bar and a coffee, Jennifer pored over the research material Kevin Debus had given her. She was reading Sciarretta's *Fundamentals of Air and Space Law* when the telephone rang.

Jennifer pounced on the receiver. "Jim?"

"*Uh... It's me, Paul.*"

"Oh. Hi."

"*I was wondering if you were OK. Will you be much longer?*"

"I don't know. I guess not."

"*Well, I'll wait up for you.*" Paul paused. "*Hey, there's a little guy who wants to talk to you.*"

"*Hi, mommy.*"

Despite herself, Jennifer smiled. "Well hello, my special little man. Have you been behaving for daddy?"

"*Yeah.*"

"Well then, how 'bout mommy brings home an ice cream from Dairy Queen for you? Wouldn't that be a nice desert?"

Dillon squealed with delight. "*Daiwy Qween!*"

In the background, Jennifer heard, "*Say thank you to mommy.*"

"*Thank you.*"

The next voice was Paul's. "*Come home soon, OK?*"

"Yeah, sure."

"*Love you.*"

Jennifer hung up in frustration. Where the hell was James? As she pondered his whereabouts, the phone rang again.

"Hello?"

"*Good evening, Ms. Simcoe.*"

Jennifer stiffened. "It's late, François. What do you want?"

"*Meet me in the lobby of the Laurier in twenty minutes.*"

"Why?"

"*It's important.*"

Her curiosity got the better of her. "Twenty minutes, then."

21:26

François was sitting on a couch in the lobby of the Château Laurier when Jennifer entered. He stood as she approached. "Thank you for coming."

Jennifer nodded.

He gestured at the door. "Let's take a walk."

They exited the hotel and made their way down Wellington Street. It was cool for a June evening, and Jennifer wished she had a jacket.

"It's funny," François began. "We've spoken to each other almost every day for the last seven months, yet we know so little about each other."

"I guess we didn't really make an effort."

"Well, it's never too late to start." He paused. "Is your home in Ottawa?"

"I live in Kanata."

"Then you're lucky. You see your family every night. I only see my family on weekends."

"…get outta here, damn Q-Bec traitors! This is *our* country…"

Jennifer turned and saw a group of teenagers gathered at the Confederation Square war memorial. Their taunts were directed at a young couple crossing Elgin Street. The man was clutching a map.

François said, "My parents used to like vacationing outside Québec. I hated it. Once I said, 'Mais maman, nous y derrions parler *leur* langue.'" He inhaled. "I remember when I was five, we went to Toronto. We went to the CNE, and I got lost in the Midway."

"That's terrible."

"I started crying, but I didn't speak English. All I could say was, 'Maman, papa, où êtes vous?' Most people ignored me. A few pointed at me, but nobody stopped to help."

Jennifer waited for him to continue.

"Then…this man and woman found me. I don't remember if they were young or old. When you're a child, every adult is just a 'big person'." He paused. "The woman knelt down and talked to me, but she spoke English and I didn't understand. The man picked me up. They took me to a place for lost children... but they didn't leave me. They stayed with me until my maman and papa came. The woman got a tissue from her purse to wipe my eyes, and the man…he even bought me ice cream."

Jennifer marvelled. "You *remember* all this?"

"I remember."

They got off Wellington, entered the South Gate of Parliament Hill, and approached the Centennial Flame. Jennifer looked up. There, high atop of the Peace Tower, was the Maple Leaf. Floodlights pointed their luminous fingers upwards, spotlighting the flag as it fluttered proudly in the dark heavens.

Jennifer caught her breath. "It's beautiful, isn't it?"

"Do you know what the ITU Convention is?"

The question snapped Jennifer back to reality. This was not a social walk; they were here on business.

"It's the constitution of the International Telecommunications Union," she recalled from Sciarretta's book, "the U.N. agency that regulates satellite communications."

"I suggest you look carefully at Article 33 of the Convention."

Jennifer was stunned. "Why are you telling me this?"

François smiled. It was the same expression she hated so much, and yet...she now saw it was sincere. Perhaps it always had been.

"Good night, Ms. Simcoe." He buttoned his jacket. "We should be going our separate ways now."

Kanata, Ontario
23:08

Paul was watching the news in the den. There was a report about a possible connection between Mehmet Ali Agca, the man who shot the Pope, and the Turkish Communist Party. He looked up as Jennifer entered the room.

Without a word, she sat beside him. Both stared at the TV for a few moments. It was Paul who broke the silence.

"I just put Dill to bed."

"Paul!" Jennifer whirled. "Why are you letting him stay up so late?"

"He was waiting for you."

"*What?*"

Paul balled his hands into fists. "He was waiting for you to bring him his ice cream from Dairy Queen! Mommy promised dessert, so Dill waited for mommy. You know how kids are! He refused to go to bed. I offered him ice cream from the fridge, but he didn't want it. Dill wanted mommy's ice cream. He just sat here...waiting. 'Mommy coming home soon,' he kept saying when I told him to go to bed. He cried a bit, but he just got too tired and fell asleep...

"We were waiting for you, Jen!"

"Don't yell at me! Why are you so upset over an ice cream?"

"It's not about a damned ice cream!" Paul exclaimed. "It's about you. It's about...us."

"You're not being reasonable! You know how important these talks. The last seven months –"

"*No!*" Paul shouted. "It's not just the last seven months. It's been from day one. I was naïve, Jen. I thought you'd change after we got married. But it got worse! Why do you always make me – make Dill – seem like a big fat zero compared to your work? Why do you find it easier to negotiate a wheat deal with the Soviets than to talk – really *talk* – to your own husband? To your family?"

Jennifer bit her lip.

"This won't end after the DNA Talks. What's next? An acid rain treaty with the Americans? A fishing dispute with Japan? Work is not your whole life, dammit! Don't we matter?"

"I'm sorry…"

"We came from different worlds, Jen. Sometimes things haven't been easy. But *both* of us have to compromise to make this work!" He pointed at himself. "I don't know, but I think I've been doing *my* part."

Langevin Block, Parliament Buildings
Ottawa, Ontario
11:52, 16 June 1981

33.2. In using frequency bands for space radio services, Members shall bear in mind that radio frequencies and the geostationary satellite orbit are limited natural resources, that they must be used efficiently and economically so that countries or groups of countries may have equitable access to both in conformity with the provisions of the Radio Regulations according to their needs and the technical facilities at their disposal.

The ITU Convention was ratified at the Málaga-Torremolinos Plenipotentiary Conference in 1973. Kevin Debus had provided Jennifer with a copy, and she spent the morning studying it. She was still unable to contact James Doherty, but another Telesat lawyer was able to give her a legal interpretation.

She began to suspect Québec wanted more than just an old satellite.

Jennifer picked up her mug. As she did so, her eyes fell upon a small photograph on her desk. The picture was taken two years ago, during a vacation to Disneyland that Paul had begged her to take. In the photo, she and Paul were standing on either side of Mickey Mouse. Paul was holding Dill. The poor kid was crying because the giant rodent terrified him.

Jennifer put the coffee down and picked up the phone. She was about to call Telesat again, but suddenly she found herself dialling another number.

"*Ottawa/Uplands Area Control Centre, good morning.*" They didn't say "bonjour" anymore.

"*Extension 293.*"

"One moment, please."

Jennifer was treated to a few seconds of canned music.

"*Terminal Control Unit.*"

"Hi, Paul."

"*Jen? What's wrong?*"

"Nothing's wrong... Uh, you're almost done for the day, right?"

"*Yes.*"

"Dill's at the sitter?"

"*Of course. Why?*"

"I was wondering... if I could see you this afternoon."

Silence.

"Paul?"

"*Jen, your negotiations are this afternoon. You've –*"

"Doesn't matter."

"*Why not?*"

"Because..." Jennifer's vision began to fog, "...it's much too nice to work."
That was their tag line when they were dating.

"*What did you want to do?*" Paul said at last.

"Anything. I just want to see you."

"*Well, Chris can give me a lift –*"

"I'll pick you up. OK?"

"*Sure. I'll wait for you by the airside security gate.*"

"I love you, Paul."

"*Never doubted that for a minute. See you soon.*" He blew her a kiss before
hanging up. Jennifer wiped her eyes before making another call.

"*Dan McTavish.*"

"Dan, it's Jennifer. Look...I can't make it to the negotiations this afternoon."

"*What?*" He lowered his voice. "*Jennifer, what's wrong. Are you sick or
something?*"

"No."

"*Then –*"

"The negotiations are off. You'd better let the Québec committee know."

"*What going on?*"

"I'm not going to be there. Is that clear?"

"*No.*"

"Well...too bad. Just tell François, OK? See you tomorrow, Dan."

"*Jen –*"

She put down the receiver.

Kanata, Ontario
7:30, 17 June 1981

Jennifer woke the instant the clock radio came on. She listened briefly to the Kim Carnes song "Bette Davis Eyes" before slapping the snooze button.

Paul was already awake. "Do you mind if I stare?" He was lying with his elbow propped on the pillow.

She kissed him. "How long have you been up?"

"Not long." He ran his hand across her shoulder. "You know, about yesterday. That was stupid, you know. I mean it was really irresponsible. Do you have any idea what you did?"

"What do you think I did?"

Paul was suddenly serious. "You put me ahead of the whole country. Two countries, I guess...I love you so much, Jen."

They drew each other close...and then the phone rang.

Paul laughed. "I was right, wasn't I? It never ends."

"Paul..."

He was still laughing. "You better get that, dear!"

With the greatest reluctance, Jennifer picked up the phone. "Hello?"

"*Jennifer!*" It was James Doherty. "*Where have you been? I've been trying to reach you since —*"

"Never mind," Jennifer snapped. "What's up?"

"*Could we meet in my office this morning? I know why Québec's so desperate for Anik A1.*"

"Something to do with its orbital slot?" Jennifer asked rhetorically.

"*How did you know?*"

"Look, time's running out. Secession day's next Wednesday. If you've got information, we need it."

"*Agreed. My office in...say, an hour?*"

"Fine." Jennifer replaced the receiver. "Paul...I'm sorry, but I've got to —"

"I know."

Jennifer gave him a quick kiss before easing herself from the warm covers.

"Jen?"

She turned.

"Give 'em hell, my love."

Telesat Headquarters
Gloucester, Ontario
9:08

"Ever heard of the Bogotá Declaration?"

Dan shook his head.

"In 1976, eight equatorial nations tried to claim sovereignty over the regions of space above their territories – up to and including geosynchronous orbit, the orbit communications satellites are stationed in. This was defeated because –"

James stopped suddenly. A trickle of blood was running down from his nose. "Aw, gee…" He whipped out his handkerchief to stanch the flow.

"Where was I? Oh, yeah." With his nose plugged, his voice sounded almost comical. "It was defeated because it violated Article 2 of the 1967 Outer Space Treaty, which states that space is not subject to national claims of sovereignty. Nevertheless, these countries had made their point."

"Which is?" Jennifer prompted.

The handkerchief was soaked with blood. "The Bogotá Declaration was a protest against perceived inequities in ITU policy." James opened a book. "The crux is the last part of Article 33 of the ITU Convention. It states that the ITU shall assign slots in geosynchronous orbit 'according to their needs…and the *technical facilities at their disposal*'. In practical terms, this means that slots are assigned in an a posteriori, or first-come, first-served basis. Many developing nations consider this discriminatory because they're afraid that by the time they get communications satellite technology, the best slots will be gone."

"So it's about orbital slots, as I suspected," Jennifer mused. "But Jim, there's something I don't understand. I spoke with one of your colleagues, and he told me Anik A1's slot wasn't that valuable. If we lose it, we can always get another one."

"It might not be that simple anymore," James said.

"Why?" Dan asked.

"The demographics of the ITU are changing. By next year, over two-thirds of the members will be developing countries, and they're taking a page from the League of Non-Aligned Nations. The next ITU Plenipotentiary Conference will be in Nairobi in a few months. These nations will unite and vote as a bloc. They'll push *hard* to get Article 33 changed."

James produced a plain brown envelope from a drawer. "I pulled a *lot* of favours for this." He struggled to extract its contents with one hand. "I've learned that the Algerian delegation will propose replacing the last part of Article 33 with the phrase, 'taking into account the special needs of the developing countries.'"

Jennifer understood the implications. "If that goes through, we might see a massive grab by those countries to take parts of geosynchronous orbit. The

ITU can't give Québec slots because it's not yet a member, and by the time it gets membership, the changes to Article 33 might be in effect. So, Québec wants to grab one of our slots before it's too late."

James nodded. "If we give Québec Anik A1, the slot automatically defaults to them."

"But Anik A1's only got a year," Dan protested. "When the satellite's gone, doesn't the slot revert to us?"

"No. The ITU will allow a country to hold a slot for up to nine years before it must be reassigned."

Jennifer asked, "Jim, what are the chances of this amendment going through?"

"Better than ever, because India's now on side. They've been sitting on the fence for years on this issue. But the ITU's just given them their slot assignments. They recently signed a contract with Arianespace to launch their INSAT constellation. Since they now have what they want, they can afford to help their allies."

"Even with India's support, those changes aren't guaranteed," Dan argued.

"Nothing's guaranteed, but with India the chances look pretty good. Québec thinks so, anyway. Remember, India's one of the largest Third World shareholders in the Intelsat consortium. They've got clout." James gingerly removed the handkerchief from his nose. The bleeding had stopped.

"Jennifer, call off today's negotiations," Dan said. "We've got to confer with Cabinet."

She nodded. "Then what?"

Château Laurier Hotel
Ottawa
13:37, 18 June 1981

Jennifer, Dan, and James worked with Cabinet through the night to draft a new deal. Now, François and Joël listened attentively as Jennifer spelled out its terms.

"Anik A1 will be *leased* to Québec for the remainder of its operational life. The price: Canada will be entitled to ninety percent of the profits from the lease of transponders to Satellite Business Systems. Any Canadian traffic currently on Anik A1, and any Québec traffic on the other satellites, shall remain in place for one year. This arrangement leaves one free C-band transponder on Anik A1, which will be ceded to Québec to do with as it chooses."

These concessions were vehemently opposed by Dan and many members of Cabinet, especially Agriculture Minister Eugene Whelan. But Acting Prime Minister Allan MacEachen decided it was the only way to end the DNA Talks quickly.

Jennifer now addressed the central issue. "At end-of-life, Anik Al will be boosted to disposal orbit...and its geostationary slot assignment shall be surrendered back to Canada. Québec will then get the transponder capacity on Anik C3 as agreed upon previously."

François rubbed his chin. "Is there anything else?"

"If Québec agrees to these terms," she continued, "Canada will support the changes to Article 33 of the ITU Convention proposed by Algeria and other developing nations."

"What?" Joël asked.

"Canada will offer its support for changes to Article 33 that are beneficial to them – if *they* support changes beneficial to us. We will propose adding the phrase, 'taking into account the geographical and geopolitical situation of particular countries.'"

François rubbed his chin. "I see..."

"Your 'geopolitical situation' is obvious," Jennifer explained, "but with our vast Arctic, the 'geographical situation' clause is good for us. These amendments will put both of us in a better position for future ITU slot assignments."

"These developing countries have become too strong," Joël said. "We believe they will force changes to Article 33 regardless of what we industrialized nations do. Why should they accommodate Canada?"

"Because we'll remind them that at the World Administrative Radio Conference four years ago, we sided with them to block the American appropriation of broadcasting services in the 12 GHz band." Jennifer grinned. "They owe us."

François leaned back in his chair. "Your offer sounds good."

"Of course, it all hinges on your government lifting its injunction against Anik D1," Dan added.

"Or else?" Joël asked.

"Or else –"

"Nothing." Jennifer cut Dan off. "Canada doesn't make threats."

François flashed his infamous smile. "Fifteen."

"Sorry?" Jennifer said.

"We want fifteen percent of the profits from the SBS lease."

"Twelve. That's final."

François drummed his fingers. "Twelve it is. The other terms are also acceptable."

Joël started, "Mais –"

"I said, the terms are acceptable." François shot his colleague a sharp glance. "Now, where do we sign?"

Jennifer and Dan couldn't contain their relief. From his perch on the wall, the framed image of Sir John A. Macdonald offered neither praise nor condemnation.

Kanata, Ontario
23:49, 13 June 1981

The final four days were hectic. After a few minor changes, the DNA Treaty was ratified. Québec lifted its injunction against Anik D1, and the satellite was trucked to the David Florida Laboratory for pre-launch testing. The secession of Québec was recognized by an all-party resolution in the Canadian parliament. With consultation from London, amendments to the British North America Act were passed, allowing separation to proceed.

Tonight, every Canadian TV channel—and one American network—was providing live coverage of the secession ceremonies in Québec City. Governor-General Edward Schreyer and Acting Prime Minister MacEachen delivered emotional speeches outside the Québec National Assembly. They praised all that the French and English had accomplished together over a hundred and fourteen years, and promised that Québec and Canada would remain friends and partners despite being apart.

Dillon squirmed on Jennifer's lap. "Mommy…"

"What is it, dear?"

"I'm tired."

"I know, sweetie. But mommy and daddy want you to watch. This is an important day, Dill. When you grow up, you'll understand."

Paul put his arm around his wife. "Poor kid's exhausted."

"It's boring…" Dillon rubbed his eyes. "It's *stoopid*…"

The family watched René Lévesque take the podium. He declared this day the most important in Québec's history since the Plains of Abraham. When he spoke of their triumph in realizing their noblest aspiration, the audience went wild.

Jennifer pulled Paul closer.

At four minutes to midnight, the crowd fell silent as the red Maple Leaf came down from the flagpole atop the National Assembly. It had been put there just for this ceremony. The camera panned the leaders. Lévesque and

Schreyer stood at respectful attention, but MacEachen looked on the verge of tears.

The blue fleur-de-lis was rising. As it climbed the pole, the crowd began to applaud. The noise crescendoed to loud cheers when the flag reached the top at midnight.

It was now the morning of 24 June 1981. The two solitudes were realized. Québec and Canada were separate.

Dillon was fast asleep.

Afterword

It is a surprising fact that the organisation responsible for regulating communications satellites predates the Space Age by almost a century. The International Telecommunication Union (ITU) was founded in 1865 as the International Telegraph Union. It took its present name in 1934 and became a specialised agency of the United Nations in 1947 as the need to regulate the use of the radio frequency (RF) spectrum became increasingly crucial after the Second World War.

Arthur C. Clarke wrote an essay in the October 1945 issue of *Wireless World* that first proposed the idea of communications satellites in geostationary Earth orbit (GEO), a prograde 24-hour circular orbit at an altitude of 36,000 km above the equator in which a satellite appears fixed in the sky from the perspective of a ground observer. The first successful GEO communications satellite, Syncom II, was launched in 1963. That same year, the ITU convened a special conference to allocate radio frequencies to the emerging space programmes of the world. By 1964, Syncom III was broadcasting the Tokyo Olympics in real-time, an accomplishment never before possible.

The ITU's mandate covers both technology and policy. Within the technical domain, the ITU promotes the development and efficient operation of space telecommunications for the benefit of the international community. In the realm of policy, the ITU becomes involved whenever an international dispute over space telecommunications arises. The agency's policies are codified in a document called the ITU Convention. Two types of meetings support the formulation of ITU policy. The first are plenipotentiary conferences where the ITU Council is elected. This council includes representation from the three ITU regions. Region 1 consists of Europe, Africa, the former Soviet Union, Mongolia, and the Middle East. Region 2 comprises the Americas including Greenland and some of the Pacific islands, and Region 3 covers the rest of the world. At these conferences, amendments to the ITU Convention are debated and adopted. Complementing the plenipotentiaries are the World Radiocommunication Conferences (WRCs) which revise the ITU's radio regulations and agree upon RF assignments and allocations.

As depicted in "Divisions", change was on the horizon in the 1970s as developing countries sought a greater influence in global affairs. Through the formation of the League of Non-Aligned Nations, developing countries became increasingly assertive in all facets of the U.N., and the ITU was no exception. By the middle of the decade, over two-thirds of the ITU membership consisted of developing countries. Through united lobbying efforts and bloc voting, they found themselves better able to challenge policies not aligned with their interests. One concern was Article 33 of the ITU Convention, which stated:

> In using frequency bands for space radio services Members shall bear in mind that radio frequencies and the geostationary satellite orbit are limited natural resources, that they must be used efficiently and economically so that countries or groups of countries may have equitable access to both in conformity with the provisions of the Radio Regulations according to their needs and the technical facilities at their disposal.

At issue was the last part of Article 33, which referred to the "technical facilities at their disposal". In practical terms, this meant that slots in GEO were assigned on a first-come, first-served basis. Many developing countries considered this discriminatory because they feared that by the time they acquired communications satellite technology the best slots would be gone. Algeria proposed replacing the last part of Article 33 with the phrase "taking into account the special needs of the developing countries". In 1976, eight equatorial nations (Brazil, Colombia, Congo, Ecuador, Indonesia, Kenya, Uganda, and Zaire) issued the Bogotá Declaration, attempting to claim sovereignty over the regions of space above their territories—up to and including geostationary orbit. Their claim had no basis in international law because it was contrary to Article 2 of the 1967 Outer Space Treaty, which states that space is not subject to national claims of sovereignty. Nevertheless, developing countries had made their point, and the amendment of Article 33 became an ITU priority. Finally, at the 1982 Plenipotentiary Conference in Nairobi, Article 33 was amended to include the compromise phrase "taking into account the geographical situations of particular countries".

In recent years, debate has emerged over how long a slot in GEO should be allowed to remain unoccupied. Some countries are pushing the ITU for increased measures to crack down on "paper satellites", or spectrum filings for satellites that are unlikely to be launched but are instead being used to hold a slot. The ITU has reduced the time between a spectrum filing and the launch of the satellite from nine years to seven years. A related aspect of the paper satellite issue is that sometimes the system deployed is not consistent with the one for which the filing was originally made. This is perhaps not surprising given the long development times for many satellite projects.

The traditional communications satellites in GEO are now being complemented by constellations of hundreds or even thousands of mass-produced small satellites in low Earth orbit (LEO). Examples include SpaceX Starlink and the planned Telesat Lightspeed. The advantages of LEO constellations include lower latency and the ability to provide telecommunications services at high northern latitudes. At the 2019 World Radiocommunication Conference, the ITU established milestones that LEO constellation operators must achieve to preserve their spectrum rights. The first satellite must be launched within seven years of the filing. Operators must then launch ten percent of their satellites within the next two years, fifty percent in five years, and one hundred percent in seven years. If an operator fails to launch enough satellites by the milestones or within the total fourteen years allotted, their spectrum rights are limited proportionally to the number launched before time ran out.

There are currently about 2,800 artificial satellites orbiting the Earth, of which approximately half are communications satellites. All of these satellites represent technological marvels and policy challenges that could scarcely have been imagined when the ITU was established in 1865.

Fixer Upper

The International Space Station was dead.

From afar, the ISS outwardly looked much as I had remembered it, a long truss structure with four pairs of massive solar arrays and the stacked cluster of dull silver and off-white pressurized modules. But its attitude, its orientation, was wrong. The main truss was tangent to the limb of the Earth, but the gradient of gravity had pulled the stacked modules into a line pointing towards the surface of its planet of origin. As we approached, I saw insulation blankets that had once been pristine and white were now cracked and stained a yellowish brown. The solar panels looked drab and were pockmarked with ragged holes in various places. There were no lights in the windows.

I had been expecting to see this, and in fact the station was in somewhat better shape than I had feared, yet I could not help but be sad. Once upon a time, a lifetime ago it seemed, the ISS had been my home in orbit for five months. To see it like this was heart-breaking.

"*Shénzhōu J-8, kāi shǐ fēi chuán duì jiē,*" intoned Shěnyáng Mission Control over the radio, indicating clearance for approach and docking.

"Shì," replied Commander Yuán Lìxúe. She turned to me. "Kristen, jì xù jiān kòng duì jiē mù biāo fāng xiàng."

As I watched the range and range-rate numbers count down on the control panel, Commander Yuán grasped her hand controller and brought our spacecraft to within two hundred metres of the sprawling, lifeless complex. The rear thrusters fired, and we begin to ease towards the docking port at a snail's pace of a few centimetres per second. At ten metres, a proximity alert flickered on the screen. I overrode the warning, and we continued forward. Lìxúe aligned the white crosshairs on the screen squarely with the alignment target on the docking port.

Our spacecraft contacted the station with the slightest bump. Nothing happened.

I glanced at Lìxúe, expecting her to call down to Shěnyáng for instructions. Instead, she simply pulled us back a few metres from the port, fired the aft thrusters again, and rammed us home harder. This time, mechanical hooks and latches swung into place and locked the vehicles together.

Yuán Lìxúe and I were the first people to visit the International Space Station in over two years.

* * *

We were docked to the Poisk research module in the former Russian segment of the International Space Station. After going through the post-docking checklist, Lìxúe and I unstrapped ourselves and floated to the hatch at the end of the Shénzhōu's orbital module.

Lìxúe asked Mission Control if they were receiving any telemetry from the station on environmental conditions. They said there was none. This was not surprising considering the station had no power, but it meant we were flying blind. There was no way to know what the environment was like on the other side of the hatch.

We adjusted our launch and entry suits, put on woollen hats, and donned oxygen masks. I threw the switch on my tank, breathed in deeply—and got nothing. The mask just collapsed around my face. I double-checked to ensure the switch was thrown. It was. I sucked in again, harder this time. The mask collapsed further around my face.

"Zěn me le?" Lìxúe asked. Her round open face was wrinkled and her black hair was streaked with grey. Unlike many Chinese her age, she did not dye her hair.

I told her my mask was not working. We went back to the Shénzhōu to look for a spare, eventually finding one in the descent module. I put on the new mask and flipped the switch. This time, oxygen flowed.

Returning to the hatch, I grasped a small star-shaped valve and turned it. Holding up a finger, I felt air rushing into the station. The pressure in the Shénzhōu began to fall. Then, the flow stopped. The ISS was still airtight.

We opened the hatch. I dove ahead, but unexpectedly bumped into Lìxúe. We looked at each other. My startled annoyance quickly gave way to embarrassment as I realized my mistake. On this expedition, I was not the commander. I gave way, then followed Lìxúe into the station.

When the air hit my face, I realized how bitterly cold it was. Moisture from my exhalations froze in a tiny cloud around my face. I played my flashlight along the darkened bulkheads of the Russian research module. The beige walls and grey electronics boxes were covered with a thin coating of ice. Mould from past occupations was frozen on the panels.

My gas detector did not register any toxic fumes. I lifted my mask and took a cautious sniff, followed by a deeper breath. The air was very cold, but it seemed to be all right.

"*Zhǐ huī yuan huì bào wēn dù?*" Mission Control wanted to know the temperature.

We looked at our thermometers, and I was surprised to discover that the scale only went to zero degrees Celsius.

Lìxúe suddenly turned, and spat on the wall. I tried to hide my disgust. She looked at her watch, timing how long it took for her saliva to freeze. Twelve seconds. "Líng xià shí dù," she declared. Minus ten degrees Celsius—a rather bracing Minnesota winter, I thought to myself.

We continued into the station, floating through the Poisk module into a small connecting node, where we did a ninety degree turn through a hatchway into the Zvezda service module. I opened the window shades to admit a little sunshine, but it was still terribly cold. With the interior lights out, the narrow fields of illumination from the windows and our flashlights created stark and eerie shadows behind the angular equipment. Sunlight streaming through the small windows lit up myriads of dancing motes and drifting rubbish. A small photograph tumbled by, and I reached out to grab it. The cherubic face of a little boy, perhaps three or four years old, smiled out at me.

Our first priority was to restore power, and with it heat, light, and full life support. Lìxúe and I removed a bulkhead panel and located Zvezda's ancient nickel–cadmium batteries, replacing them with modern solid-state electrolytes. Connecting the new batteries to Zvezda's power bus was a difficult task because we had to take off our gloves, and our hands soon become painfully cold and stiff. The silence was oppressive. With no motors or ventilators whirring, this frozen nook in space was the quietest place I had ever been.

My oxygen mask suddenly collapsed against my face. Startled, I looked down at the tank and saw the display still showing around forty percent. I tapped the tank and flipped the switch back and forth, but no more air flowed. Resigned, I took off the mask and continued working. But without ventilators to circulate the air, exhaled carbon dioxide hovered about my face.

My head began to ache, my arms and legs grew sluggish, and I started feeling drowsy.

"Nǐ méi shì ba?" Lìxúe said, asking if I was all right. Seeing my mask off, she handed me hers.

I brusquely told her I was fine. It really irritated me, their attitude that Westerners in general and Americans in particular were weak and needed looking after.

The ISS creaked and groaned under thermal stresses as it passed from sunlight into the Earth's shadow. Lìxúe and I retreated back to the relative warmth of the Shénzhōu to ride out the orbital night. Forty-five minutes later, we emerged to resume our work.

The first battery was hooked up, and I saw Lìxúe smile as the voltage rose. The job went quicker for the other seven units. After a final check of the connections, I switched on the main power. Suddenly, the dead module sprang to life. Fans began to whir and the ventilation system kicked in with a low hum. Displays illuminated on several pieces of equipment. The interior lights came on.

"Shěnyáng, diàn lì huī fù chéng gōng!" Lìxúe smiled broadly as she reported our success to Mission Control.

A moment later, everything went dead again.

* * *

We eventually managed to restore partial main power. This took the better part of a day, by the end of which we were simply exhausted.

But there was one more obligation. Shěnyáng Mission Control told us to standby for a media event in twenty minutes. In my former life at NASA, press conferences were actually something I enjoyed. Outreach and education had been important parts of my job, and I relished sharing the wonders of spaceflight with the general public.

Lìxúe unrolled her tablet and set up a camera and lights on the bulkhead. She said I looked tired and told me to let her do the talking.

I nodded curtly. Because nobody is interested in what the dumb Red Card American has to say, I thought to myself.

"Huānyíng lái dào CCTV," the interviewer began. "Wǒ shì zhǔ chí rén Dù Tíngfāng. Jīn wǎn, wǒmen... chǎng...jiǎng tàikorén Yuán Lìxúe hé Kristen Bartlett..."

The rapidly spoken Chinese, exacerbated by the poor quality communications link, was hard for me to follow. Lìxúe spouted platitudes about the importance of our mission, how the station was in great shape (even though we had only been here two days and had not yet ventured into most of the other modules), and how much she enjoyed working with me (even though the feeling was not necessarily mutual). The interview lasted about fifteen minutes, but it was rough because the comms kept dropping out. Through it all, I stayed quiet with a forced smile plastered on my face. There were no questions for me.

Finally, the interviewer signed off. My fatigue returned with a vengeance, and I was expecting to head back to the Shénzhōu for some much needed sleep when Mission Control called up again. We were told to standby for a message from Liǔ Diānrén, CEO of the Xīn Shìjiè Corporation and the financial backer of our expedition,

Dismayed, I turned to Lìxúe. She didn't appear surprised.

Liǔ Diānrén appeared on the screen. He looked to be in his late thirties with slicked black hair, a pale square face, and an angular beard. Our sponsor was speaking from his office at Xīn Shìjiè's headquarters in Sharjah. I found his Chinese even harder to follow than the CCTV host. He congratulated me and Lìxúe for a job well done, and reiterated his dream of bringing the station back to life and creating a new heavenly dynasty or words to that effect. By the time he started babbling about something called Cháng'é Bèn Yuè, he had pretty much lost me. But I caught the last bit, in which he announced that henceforth the module we had successfully reactivated would be known by its new, proper Chinese name—Dōngxīng, the Eastern Star.

Strangely, we didn't lose comms once during the entire rambling monologue. After what seemed an eternity, Liǔ Diānrén finally shut up and the screen went dark.

Within a few seconds, so did the rest of the "Dōngxīng" module.

* * *

Dōngxīng was only the start of the name game. As Lìxúe and I reactivated modules, Liǔ Diānrén called up to personally rechristen each with Chinese monikers. The Zarya functional cargo block became Shǔguāng ("New Epoch"), the Unity node was now Níngjìng ("Serenity") and the former U.S. laboratory module Destiny became Wángcháo ("Dynasty"). I found this creeping Sinofication—a cultural appropriation as much as a technological one—quite upsetting, all the more so with the knowledge that I was abetting it.

Power, life support, and communications had been more or less stable over the past week, enabling Lìxúe and me to prepare for the arrival of the last three members of our expedition. The lights were dim and the station smelled like a musty old wine cellar, but we tried our best to clean up. We stuffed excess gear and broken-down equipment behind panels, mopped up globs of floating water, scrubbed down the bulkheads with fungicide wipes to remove the mould, and installed new filters in the air cleaning system. It was a far cry from the ISS that I remembered, but things were at least in better shape than what had greeted me and Lìxúe just over a week ago.

From a scarred window in the Dōngxīng service module, I watched the approach of the Shénzhōu J-9 spacecraft to the docking port of the Kēxué laboratory. A momentary shudder reverberated through the station at the moment of contact.

I swam into the Kēxué module. Lìxúe opened the hatch, and the final three members of our expedition emerged.

"Zīchéng! Wénxìn! Chéngfēi!" I called out.

"Kristen!"

The commander drifted aside, apparently allowing me to greet the newcomers first. Fàn Zīchéng was a burly giant of a man with jet-black crew cut hair. It always amazed me that he was able to fit his immense frame into the cramped Shénzhōu. Next was Cài Wénxìn, a greying bespectacled man who in every way appeared the physical opposite of Zīchéng. Last was Zhāng Chéngfēi, a middle-aged man of medium build with a round, puffy face. I had worked with these guys in Shěnyáng for months with the original mission commander, whom Lìxúe had replaced in the final weeks of our training.

They each gave me a hug as they exited the Shénzhōu, before floating over to Lìxúe to shake her hand and exchange a few words. We all then migrated in single file—heads closely following feet—into the Wángcháo laboratory, where we alighted to endure another video message from Liǔ Diānrén.

When this guy talked the comms never failed, much to my disappointment. Liǔ was either speaking slower or the long solitary confinement with Lìxúe had improved my Chinese, because I actually made out most of his rambling monologue.

"*First you were two, and now you are five. I wish you were eight, for that would be of good fortune, but my engineers tell me the station cannot sustain more than five. Perhaps I will fire them! But five it is now, and you must work together as one, like the white shadow that moves unseen across the heavens. You must succeed, for if you do not, the fault will be yours. Work well, you five heavenly shadows!*"

I blinked and shook my head. Perhaps my Chinese hadn't improved as much as I thought, because nothing Liǔ said made sense to me. I glanced over at Zīchéng, Wénxìn, and Chéngfēi. The three guys looked excited, occasionally exchanging quiet words and pointing at the screen and nodding. Lìxúe just stared ahead with her arms crossed, her face an expressionless mask.

* * *

Docked to the aft port of the Dōngxīng service module was Progress MS2-1C, the last in a series of Russian expendable cargo spacecraft that had delivered supplies to space stations since the late 1970s. This particular vehicle had been heavily modified with a larger service module type engine, extra fuel tanks, and the ability to draw residual propellant from the station itself. Its mission had been to deorbit the International Space Station following its abandonment. But in a final act of reckless optimism, the original international partners agreed instead to use the Progress to boost the ISS into a higher orbit, with the hope that a future crew might someday reactivate the station.

It had always been my dream to command such a crew and revive the ISS. Not being in charge was bad enough, but being the minion of a Chinese expedition was something I never would have guessed and don't think I've quite gotten over.

Lìxúe and I were loading broken equipment and garbage into the cargo module of the Progress. It was not a pleasant task. We found ourselves entangled in a chaotic cloud of rubbish, trying to push the floating garbage bags and obsolete electronics into the Progress. I gagged from the smell of decomposing trash.

There had never been much small talk between me and Lìxúe, both because of the language barrier and also the fact that I really didn't know her well. But the cold, damp, and malodourous Progress was just too much. I needed a distraction.

"What you know of Liǔ Diānrén?" I said at last in my awkward Chinese.

Lìxúe stopped, a garbage bag frozen in mid throw. I feared she might have thought my question impertinent or disrespectful. But then she spoke.

"Liǔ Diānrén lives in a cloud of success. But it's not his success." She tossed the bag into the Progress.

There was an awkward silence. Her answer was curious, and now I wanted to hear more. "Explain."

Lìxúe hesitated. A strange look crossed her face. Perhaps she thought she'd already said too much, but she continued. "Diānrén's father, Liǔ Fākuàng, had built up the Xīn Shìjiè Corporation from a dumpling stall in Xī'ān into the diversified global corporation it is today. Liǔ Fākuàng was lucky enough to strike gold, but his son wasn't there when he was swinging the pick. Diānrén merely inherited the throne of his father's industrial empire when the elder Liǔ became ill and retired early, a dynastic succession if you will."

I didn't know any of this, and I was fascinated.

"Do you know what is a shǎodì?" Lìxúe asked.

"Little emperor?" I repeated.

"Liǔ Diānrén is like me, an only child – an only son – born of the first generation under the old One-Child Policy. The difference is, my family wasn't rich." Lìxúe paused. "He has a full-time employee whose only job is to peel grapes for him, the way his grandmother used to when he was small."

And I thought my daughter was a princess.

Lìxúe sighed. "We are a nation of badly brought-up children."

The Progress cargo module was nearly filled to the brim. After checking that the seal around the hatch was airtight, we loosened the bolts that secured the Progress to the docking ring of the Dōngxīng module.

"Shěnyáng, we are go for Progress separation," Lìxúe radioed down.

"*Chéng rèn,*" Mission Control acknowledged.

Ground controllers commanded latches to open, and springs pushed the Progress away from the station. Floating to one of the small scratched up windows in Dōngxīng, I watched the vehicle back away, plumes of thruster exhaust periodically puffing from the squat insect-like spacecraft. I continued to watch until it resembled nothing more than a distant star in the blackness of space, destined for a fiery re-entry over the South Pacific.

Good riddance to bad rubbish.

* * *

It took weeks to restore sufficient functionality to the Space Station Remote Manipulator System, the seventeen metre long robotic arm that had been crucial to the original construction and maintenance of the ISS. Most of the Canadian engineers who had developed the SSRMS were either retired or dead, but Liǔ Diānrén's team managed to find a handful of the old timers and brought them to Shěnyáng. Despite their best efforts, only one of the arm's redundant control strings would respond to commands, and the wrist roll joint had failed completely. We would also have to make do with the arm's remaining 386-processor, an archaic chip that had been obsolete even when it was originally launched.

"*Confirm Dàshǒubì is in the pre-capture position,*" Shěnyáng called up, referring to the robotic arm by its new Chinese name.

"Acknowledged." I looked out the Cupola windows, pitted and pockmarked from years of dust and debris impacts. Above the hazy blue of the Earth's limb, the arm hung in space like a giant articulated soda straw, its formerly white thermal blankets now a dull yellow from long exposure to ultraviolet radiation and atomic oxygen. About ten metres below was Huǒniǎo—the Phoenix. Twenty metres long and five metres wide, Huǒniǎo was a cargo vehicle made up of half the payload shroud of a Chángzhēng-5G heavy-lift rocket with the modified upper stage still attached as a propulsion module. Resembling a truncated version of the old Space Shuttle payload bay, it was packed with tons of hardware, equipment, and tools we needed to complete the station's reactivation.

"Start capture." My hands tightened over the hand controllers of the robotic workstation. I had operated the robotic arm on my first expedition to the ISS. They say that one never forgets how to ride a bicycle, except in this case the bike had one flat tire and a messed up gear shift. The arm was sluggish and I had to fly a non-optimal trajectory to compensate for the failed wrist roll joint, but eventually I managed to snare the Huǒniǎo's grapple fixture. I then drew the cargo vehicle closer to the station and manoeuvred it to the berthing mechanism on the Héxíe ("Harmony") node.

"We have Huǒniǎo," I reported.

"*Gàn dé hao*," Shěnyáng radioed.

"Hěn hǎo!" Lìxúe exclaimed happily. "Well done indeed!"

With the capture complete, I cycled through the robotic arm's cameras trying to find a working unit. One of the boom cameras eventually responded to commands. Through a grimy lens obscured by years of brownish-yellow propellant residue, I surveyed the berthed cargo ship. Most of the payloads were familiar to me: the new control moment gyroscopes, the high-pressure gas tanks for the airlock, the ammonia coolant tanks, the batteries, and…

I frowned. There was cargo I did not recognize from the manifest: folded-up structural members, and oddly shaped tanks of unknown content. I powered down the robotic arm and went in search of Lìxúe to inquire.

The commander was in the Wángcháo laboratory module, hunched over an open avionics rack like a surgeon at work. She looked up from the exposed spaghetti-like mess of wire bundles and electronic components.

"Kristen, good job with the Huǒniǎo," Lìxúe said.

"Xiè xiè." I paused to compose the sentence in my head. "Lìxúe, I see cargo in Huǒniǎo not known from manifest."

Lìxúe looked at me with an odd expression. After a moment, she said, "I will brief the crew after dinner."

* * *

Meals are of great cultural significance to the Chinese, in which the social interaction is far more important than the eating. I was certainly fine with the latter, but my lack of linguistic fluency was still a barrier to the former. This was the case during our training on the ground, and nothing had changed in space. So, once again I ate dinner in silence while Zīchéng, Wénxìn, and Chéngfēi chatted amongst themselves.

The guys looked terrible. Wénxìn and Zīchéng had been trying to fix a coolant leak in the Dōngxīng service module and must have gotten some fluid on their faces because their eyes were red and appeared the size of golf balls. Chéngfēi had been trying to replace a fan and filter in the Kēxué multi-purpose laboratory module and had scratched his arm on something while reaching behind a panel. It looked swollen and infected, a dull shade of blue from wrist to elbow like a botched ink job from a shady tattoo parlour.

As I ate my can of lukewarm chicken sticky rice and tube of watery soya milk drink, I tried to follow their conversation as best I could. Wénxìn seemed to be mocking Yuán Lìxúe's son, who was apparently a thirty-something slacker still living at home with his biological father, fantasizing about becoming a big-shot business person like Liǔ Diānrén but never actually bothering to do anything that might realize such a dream. Chéngfēi suggested maybe the guy should join the PLA, to which Zīchéng replied the only army that would take such a loser was the Americans.

They laughed. I pretended not to understand.

Lìxúe floated into the module, and the guys quickly fell silent.

"Nǐ chīfàn ma?" I said, asking Lìxúe if she had eaten.

Lìxúe shook her head. "No, but thanks for asking. I've been busy." She turned to the three guys. "If you're all finished eating, I need to brief you on a significant change to the mission plan."

Lìxúe unrolled her tablet and mounted it to the bulkhead. "Now that the final hardware elements have been delivered by the Huǒniǎo vehicle, I can tell you the full scope of our mission." She woke up the tablet. "Liǔ Diānrén has a great dream. His ambition goes beyond this station endlessly circling our home planet. He wants to push humanity out to a new horizon again. Liǔ Diānrén wants to send this space station to the Moon."

My jaw dropped. The faces my companions registered similar expressions of surprise.

The tablet showed a simulation of the station's orbit around the Earth. "In three weeks," Lìxúe continued, "a cryogenic upper stage will dock with the station. Following our departure, this upper stage will perform a series of propulsive burns at perigee over the course of a month." On the screen, the station's circular orbit began to stretch out into an ellipse that eventually reached out to lunar orbit. "This will put the station into an Earth-Moon cycling orbit, one that will bring the station into nodal alignment to pass behind the Moon every other orbit, about twice a month. Once established in EMCO, an Advanced Shénzhōu vehicle will be launched to deliver the first lunar mission crew to the station."

We stared in stunned silence. Finally Wénxìn said, apparently for lack of a better question, "This new mission…does it have a name?"

"Wàn Hù," Lìxúe replied.

The three guys looked puzzled, but I recognized the name. Wàn Hù was a Ming Dynasty official. According to legend, he was the world's first tàikōnaut, building a chair with forty-seven rockets in an attempt to launch himself into space.

"Any more questions?" Lìxúe asked.

There was something about the rocket-chair story that bothered me, and at last I remembered. Wàn Hù had blown himself up.

* * *

With a full crew of five and the tons of hardware brought up by Huǒniǎo, the pace and scale of activities picked up significantly. Over the course of two weeks Wénxìn, Zīchéng, and Chéngfēi paired up in turns to conduct a series of demanding long-duration spacewalks, or EVAs. The first excursions replaced the four failed control moment gyroscopes on the Z1 truss. Once activated, the new CMGs restored full attitude control and enabled the station to overcome the gravity gradient torque that had pulled its axis of modules downward, returning the station to its nominal local-vertical local-horizontal orientation with respect to the Earth's surface. Subsequent EVAs installed structural reinforcements at various locations about the station's exterior: the connection between the Héxíe node and the Shènglì (formerly Kibo) laboratory, as well as the connections between the Shǔguāng functional cargo block, the pressurized mating adapter, and the Níngjìng node.

I was worried for the guys because even in the best of circumstances EVAs were not easy. Indeed, they could be positively dangerous. During the original construction of the ISS, astronauts and cosmonauts would have trained for months if not years to perform complex assembly tasks like these. We prepared instead by studying written procedures and watching videos of practice sessions conducted in the neutral buoyancy water pool in Shěnyáng. It was hardly ideal, to say the least.

A few days later, it was my turn.

The airlock in the former U.S. segment of the station had been called Quest but was now renamed Xīngmén, which agreeably translated into English as "Stargate". The airlock compartment was worn and scruffy. On a wall was a faded mission patch sticker with some Russian words scrawled underneath.

We put on our helmets and donned our gloves. Lìxúe started the depressurization pump, and the air bled away. I was nervous.

"*Lìxúe and Kristen, you are go for hatch opening,*" said Mission Control.

"Hǎo ma?" Lìxúe asked.

I took a deep breath, and nodded.

Lìxúe vented the residual air before releasing the handle and opening the hatch.

The Sun was rising. Our EVA was planned to start at orbital sunrise so that we would get the longest period of light. The entire planet was spread out beneath us like a giant blue, green, and white tapestry.

"Wǒmen zǒu ba," Lìxúe said.

We tethered ourselves to the outside of the airlock and performed a final inspection of our tool and equipment packages as well as our Jīnyì propulsion units. After executing the self-test, the intention light on my Jīnyì went green. Lìxúe's, however, stayed red. She tapped it with a gloved hand until it turned the right colour.

I could see faint puffs from Lìxúe backpack as I followed her along the length of the Wángcháo laboratory module to the Huǒniǎo cargo ship berthed at the Héxíe node. Our job was to transfer the conformal water tanks that had been brought up in the Huǒniǎo and install them on the outside of the Níngjìng node. Arriving at the Huǒniǎo, Lìxúe and I un-holstered our EVA power tools and set to work. We applied our wrist tethers to eye hooks on the first water tank before releasing the tie-rod bolts with our power tools. Then, we grabbed the tank by handles at each end and pulled the vessel free of its flight support equipment.

"Shěnyáng, Kristen and I have the first tank," Lìxúe reported.

"*Chéng rèn,*" Mission Control acknowledged.

Holding the tank between us, Lìxúe and I flew back along the Wángcháo laboratory to the Níngjìng node, where the Xīngmén airlock was also located. During an earlier EVA, Wénxìn and Zīchéng had removed some of the micrometeoroid and orbital debris panels, exposing the structural attachment points underneath. Using our power tools, Lìxúe and I bolted down the first tank, which was shaped to fit the cylindrical exterior of the node.

After another seven exhausting hours, Lìxúe and I had installed the remaining water tanks onto the Níngjìng node. We then connected the station's fluid lines to the tanks and also installed an external radiation dosimeter. Our tasks completed, I started making my way back to the Xīngmén airlock.

Lìxúe was not following me. Puzzled, I turned. "Everything good?"

"Yes Kristen, everything's fine," Lìxúe said. "I just have a small task to perform."

Lìxúe pulled a small octagonal object out of a pouch. It was a bāguà mirror. According to Chinese superstition, a bāguà is supposed to deflect bad qì from malevolent outside forces. I thought to myself that, really, there is no qì in space—good or bad—because qì literally means "air".

Lìxúe screwed the bāguà into an unused tie-rod bolt hole on one of the water tanks. She jiggled the bāguà to make sure it was secure, then pulled out two more objects—a decal of a Chinese flag, and one of the Xīn Shìjiè Corporation's crescent moon logo. She stuck both proudly to either side of the bāguà.

"*The most important task of the EVA has been completed*," said Mission Control, with no obvious intention to be ironic.

There was the sound of a tiny explosion, like a kid's cap gun going off. A spider-vein crack suddenly appeared on my faceplate.

"Shit!" I screamed in English, instinctively putting my hands to my face.

"Kristen!" Lìxúe was with me in seconds, her hands reaching over my suit. She quickly found the control for the secondary reserve oxygen tank and cranked up the flow.

"*What happened?*" Shěnyáng demanded. "*What's going on?*"

"Emergency ingress!" Lìxúe declared. "Kristen's helmet is breached."

"*Get back inside immediately*," Mission Control ordered. "*Her suit pressure is down to 30.6 kilopascals and falling.*"

"Kristen, listen to me," Lìxúe said calmly. "I need you to take your hands off your faceplate. Please, just for a moment."

Reluctantly, I complied. Lìxúe slapped a strip of silvery-grey material to my faceplate. It took me a moment to recognize the stuff.

Duct tape.

"Put your hands back now," Lìxúe said. As I did so, she grabbed my arm and activated the thrusters on her Jīnyì jetpack to push us the last couple of metres towards the Xīngmén airlock. She opened the hatch and unceremoniously dumped me inside.

"Are you all right?" Chéngfēi asked over the intercom.

"Yes, we're fine," Lìxúe replied.

When the airlock re-pressurized, I removed my helmet and took a deep breath. The station had a faintly unpleasant odour, something between gasoline and antifreeze. Lìxúe and I looked at each other. Our sense of smell must have been deadened by the weeks of exposure to the sketchy air.

"Smells like Běijīng," Lìxúe said, and then smiled.

I shook my head and chuckled.

Yuán Lìxúe brought duct tape on a spacewalk. How cool is that?

* * *

On Chinese New Year's Day, we un-berthed the Huǒniǎo cargo ship from the Héxíe node and released it into space. Under command from Shěnyáng Mission Control, Huǒniǎo fired its thrusters and pulled away, heading for a fiery plunge into the South Pacific.

That afternoon, Liǔ Diānrén announced through a press release that the reactivated station was now fully operational. Officially. I supposed it was mostly true. Liǔ declared a holiday for the crew and ordered a celebratory meal, both for New Year's and also as a send-off for the impending departure of Fàn Zīchéng, Cài Wénxìn, and Zhāng Chéngfēi.

We set up a dining table in the Níngjìng node, allegedly the perfect location for the New Year's dinner because it was now supposedly shielded by the bāguà against bad qì. Against a bulkhead the colour of bad teeth, Chéngfēi and Zīchéng put up a red banner with the characters 完成任务—"mission accomplished".

Much like the rest of the complex, the node was a chilly, dimly lit place. The station was simply underpowered, its solar arrays now providing only a fraction of the power they had generated when new, their photovoltaic cells degraded after decades in orbit. Our installation of newer, more energy efficient electronics and lighting systems offset some of the losses, as did leaving unpowered the modules of the station that we were not be using, but it wasn't really enough.

Once again, our meal could not proceed without the annoyance of one final video message from Liǔ Diānrén.

"*First, there was a dream,*" Liǔ intoned solemnly. "*Now, there is reality. Once, it was called the International Space Station. The name itself was a cruel, humiliating joke. How could the station have been 'international' when China was deliberately excluded, despite the fact that even then we had a space program at least equal to that of the United States?*"

"*No more humiliation! Today, we Chinese have made another great leap forward into the untainted cradle of the heavens. The guǐlǎo have abandoned the stars. Let them now look up and pay deference to the ultimate dynasty that I have created, and know there is a new order in the heavens. Henceforth, it will no longer be the 'International' Space Station but will be known as Xīn Shìjiè Tiān Zhōu, a new spacecraft for a new world, my gift to humanity, a vessel that will soon go even farther than anyone has dreamed.*"

Zīchéng, Wénxìn and Chéngfēi smiled broadly and patted each other on the back. Lìxúe and I watched in stony silence. With his slicked black hair, demonic beard and ugly ass yellow-beige suit, I thought Liǔ looked like a James Bond villain. The guy was crazy.

"*Gōng xǐ fā cái, my loyal comrades, and enjoy the New Year's feast. Eat lots of food to build up your strength so that you can work even harder next year!*"

We dug into our meal as soon as the video link closed. On paper, it looked like a feast, an eight course meal starting with tubes of swallow's nest soup and tins of jellyfish salad, followed by thermo-stabilized duck, fish and dumplings, rehydrated báicài and mushroom noodles, and finishing with mango pudding cups. In practice, everything was lukewarm except the dessert, which was cold and watery. Our food had been like this throughout the mission, but I guess I was expecting better for New Year's.

After dinner, Yuán Lìxúe handed out small red gift bags for each of us. Inside was a cheesy booklet commemorating the Xīn Shìjiè Corporation's 25th anniversary, a chocolate coin in gold foil wrapping, a pair of red socks, and most surprising of all, a mandarin orange.

"Lìxúe, this is unbelievable!" Wénxìn stammered. "Where were you hiding these?"

I put the mandarin to my nose and inhaled deeply. It was not fresh and the smell was faint, but the scent of fruit from the green Earth almost overwhelmed me.

"Xiè xie, Lìxúe," Chéngfēi said with heartfelt sincerity. "I can't think of a better way to end our time together in space."

Three days later, Zīchéng, Wénxìn, and Chéngfēi donned their launch and entry suits in preparation for departure. Following established procedure, they would take the older Shénzhōu J-8 spacecraft back to Earth, leaving the newer Shénzhōu J-9 vehicle for me and Lìxúe. The Zhìhuì research module to which Shénzhōu J-8 was docked was cramped for five, so the goodbyes were brief.

"You are hereby relieved of your duties," Lìxúe said.

Wénxìn turned briefly to Chéngfēi and Zīchéng before speaking. "We stand relieved."

Lìxúe and I embraced each of the guys as they passed, exchanging expressions of thanks and goodwill. The three of them floated in turn through the passageway into the Shénzhōu, and the hatch closed behind them.

An hour later, Shénzhōu J-8 disengaged from the Zhìhuì docking port. I watched from the scarred Cupola windows, the rugged peaks of the Peruvian Andes providing a spectacular background to the retreating insect-like spacecraft. I continued to watch until the Shénzhōu was only a point in the distance, hurtling towards a touchdown at the Sìzìwáng landing site in Inner Mongolia.

* * *

Our final visitor, the Wàn Hù lunar boost vehicle, arrived a week later. Lìxúe and I monitored its approach from the Cupola. I cycled through the station's external cameras, but only the unit at the end of the P1 truss worked well enough to provide an image. Something was amiss. The camera zoom settings were much lower than I had expected. I looked out a blurry window. My eyes widened as it got closer and its size became apparent. The Sun momentarily passed behind the booster, and we were briefly eclipsed in darkness.

"Huge," I whispered in awe.

"Yes," Lìxúe said.

Now within a few metres of the Dōngxīng docking port, I got a good look at the monstrosity. It was cylindrical, almost thirty metres long and eight metres in diameter, and covered with rusty orange insulating foam. A large radiator panel jutted out from one side. At the nose were rendezvous sensors and a docking mechanism, and at the back was a module with a pair of large engine nozzles surrounding by quads of smaller thrusters.

The Wàn Hù booster ploughed into the back of the Dōngxīng service module. A shudder reverberated through the station. From the Cupola windows, I saw the station's central axis of pressurized modules writhe under torsion like an awakening dragon. Wàn Hù was more than twice the size of any of the station's modules.

"Shěnyáng, we have contact and capture," Lìxúe reported.

We spent the next three days preparing the station for its second, and hopefully much briefer, dormancy. Lìxúe and I closed many of the interior hatches to prevent the whole station from depressurizing in case one of the modules suffered a debris strike. The common cabin air assembly—basically the station's dehumidifier—was cranked up to control moisture and reduce microbial growth. We shut off as many pieces of equipment as we could to lower the risk of fire, but left the water processor assembly active to keep fluids flowing through the lines and prevent stagnation.

The first engine burns to inject the station into the Earth-Moon cycling orbit were to start at the end of the third day, before cryogenic boil-off became a problem. I made my way to the Dōngxīng module and donned my launch and entry suit, then went to meet Lìxúe in the Kēxué laboratory where the Shénzhōu J-9 spacecraft was docked. She was already there, waiting for me, still dressed in a T-shirt, long pants…and a pair of bright red socks.

I understood immediately.

"You not coming," I said.

Lìxúe shook her head.

"Why" I asked.

"I am to stay aboard and monitor the EMCO insertion," Lìxúe replied, "and intervene if anything goes wrong."

"Crazy," I said.

"I'll be fine," Lìxúe said stoically. "Life support is still sized for five people, and there's lots of food and water. Once the station is established in EMCO, the Advanced Shénzhōu will arrive with the lunar mission crew...and I will be here, waiting for them."

"Many journeys through Van Allen radiation belts," I said quietly.

"The transits are relatively brief, and the modified Níngjìng node is an adequate radiation shelter."

"Why you really do this?" I asked.

"My family...my son. We need the money."

I nodded, understanding. I had needed the money too. It really was that banal.

"It's ironic, you know," Lìxúe said. "The name, Wàn Hù. It's not Chinese."

I raised my eyebrows.

"The story never appears in historic Chinese literature," Lìxúe continued. "The first mention was actually in an American magazine, I think in the early 1900s. It's a nice story, but it's not Chinese."

I pondered that for a moment, then said, "I am relieved of duties."

"You are relieved," Lìxúe said.

We stared awkwardly at each over for a moment, then hugged. The embrace was warm and sustained. "Zài jiàn, hǎo péngyou," I whispered.

"Time to go," Lìxúe said.

I saluted Lìxúe, then dove into the hatchway to the Shénzhōu. Inside the orbital module of the cramped spacecraft, I turned. Lìxúe was looking at me. The round hatch began to close, gradually eclipsing her face until it was gone.

* * *

There was the slightest thump as the holding latches and hooks were released and springs pushed Shénzhōu J-9 away from the docking port.

"Shěnyáng, separation at 11:38," I reported.

The station looked vastly different than what Lìxúe and I had seen when we arrived almost three months ago. Most prominent was the bloated orange bulk of the Wàn Hù docked at the end of the Zvezda service module. The massive Chinese booster was almost two-thirds the length of the station itself, from Zvezda to the Harmony node. Along the main truss, the solar arrays and radiator panels were retracted in preparation for the upcoming propulsive thrust, the largest ever attempted in space.

Alone in the Shénzhōu, I found myself referring once again to the station modules by their original names.

I was now above and behind the ISS. My video survey complete, I fired the thrusters again, ending the fly-around and breaking away from the station's vicinity. In a higher, slower orbit, I watched the massive complex pull away. Squinting at the screen, I thought I saw a small panel fall away from the station.

Half an hour later, the ISS was five kilometres ahead of me, soaring gracefully over the Taiwan Strait.

Over the radio, the commentator in Mission Control was counting down. "...sān...èr... yī...fā shè!"

The massive cryogenic engines of the Wàn Hù booster ignited.

"Shùn fēng, Yuán Lìxúe," I whispered. "You're going to the Moon."

"*Wàn Hù's two main engines are at maximum thrust,*" Mission Control reported.

Zooming the Shénzhōu's camera on the accelerating ISS, I saw a pair of fuzzy, incandescent auras streaming from the twin nozzles of the Wàn Hù booster. There was little sense of speed, but this first engine firing was supposed to increase the station's velocity by about two hundred and seventy metres per second.

I took a quick glance at my watch. Seventy seconds had elapsed in the planned three minute burn.

Four seconds later, it all went wrong.

At first, it was difficult to see what was happening. I blinked, not quite believing my eyes that were telling me the central axis of the space station's modules no longer appeared quite straight. The calamity became obvious a fraction of a second later when it visibly distorted into a flattened V-shape, the vertex of the bend near the intersection of the module stack and the transverse truss. With a painful knot forming in the pit of my stomach, I realized that I was witnessing a catastrophic structural failure at the junction between the pressurized mating adapter and the former Unity node.

"Oh, crap!" I exclaimed in English. "Sweet mother of –"

The Wàn Hù booster was still firing as the structural failures cascaded. The line of pressurized modules was now severed behind the old Unity node, like a string of sausages cut by a butcher. Seconds later, the central truss snapped like a twig between the S0 and P1 elements. Then, the former Japanese module Kibo broke away from the Harmony node.

"*There appears to have been a major malfunction,*" intoned the cold voice of the Mission Control commentator. "*Telemetry has been lost.*"

I keyed the radio. "Shěnyáng! Terrible structural failure!" Without waiting for a response, I tried to raise the commander. "Yuán Lìxúe, respond!"

Starved of propellant, Wàn Hù's engines finally went dark.

The space station was shattered into five major pieces, with dozens of smaller bits floating about. Clouds of gas, frozen fluid drops and shards of equipment briefly spewed from ruptured modules until the air was completely dispersed to vacuum. Amongst the debris was a small octagonal mirror—the bāguà, which proved itself completely useless in warding off disaster. In time, the smashed remnants of the station would succumb to orbital decay, bringing down to Earth the mad dreams of a little emperor in Sharjah.

The International Space Station was dead.

Afterword

In the final episode of the 1990s science fiction TV series *Babylon 5*, the titular space station is decommissioned by deliberately overloading its fusion reactors and blowing the place to smithereens. "We can't just leave it here, it would be a menace to navigation," an Earthforce commander tells former president John Sheridan, saying the station had "become sort of redundant" and citing recent budget cutbacks. This is a peculiar action because one would think a massive cloud of debris in the Epsilon Eridani system would be an even greater menace to navigation. A more logical decommissioning would have been to crash the station onto Epsilon 3, the planet about which it had orbited, although I suppose Draal and the Great Machine might have taken offense.

The first element of the International Space Station (ISS), the Russian-built Zarya control module, was launched in November 1998. More than two decades later the refurbished backup of the Zarya module, renamed the Nauka multipurpose laboratory module, was launched and attached to the ISS in July 2021. Another new Russian module, a five-port docking node called Prichal, was attached to the nadir (Earth-facing) side of Nauka in November 2021. NASA is also planning to expand the ISS. In January 2020, the U.S. space agency awarded an initial contract to a private company called Axiom Space to provide up to three new commercial modules to be attached to the ISS, the first of which is scheduled for launch in 2024.

The current multinational ISS partnership—consisting of the United States, Russia, the European Space Agency, Japan, and Canada—is discussing the extension of ISS operations to 2030 and perhaps even longer, but like its fictional counterpart on *Babylon 5* at some point its mission will end and the station will have to be removed from orbit. ISS will be the largest human-made object ever brought down to Earth, and doing so safely will be a challenging engineering problem. Nobody wants a repeat of the (mostly) uncontrolled 1979 re-entry of Skylab, in which parts of the station ended up hitting a remote region of Australia and resulting in a $400 littering fine to NASA.

One of the cargo vehicles that brings supplies to the ISS is called Progress, a type of spacecraft the Russians have been using since the late 1970s to support their earlier generations of space stations. The current ISS decommissioning plan envisions using a modified Progress equipped with a more powerful engine, extra propellant tanks, and the plumbing needed to use up any remaining fuel in the Russian segment of the ISS. This modified Progress would execute a series of high-thrust burns to lower the ISS orbit and eventually bring the station to a fiery destructive plunge into a remote part of the ocean—likely the South Pacific or the Indian Ocean—as far away from major shipping lanes as possible (see "menace to navigation").

Such was also the fate of the immediate predecessor to the ISS, the much-maligned Russian space station Mir. For a brief time, the Mir actually got a temporary stay of execution. In 2000, a private firm called MirCorp signed an agreement with the Russian space company RSC Energia to lease the station for commercial activities. MirCorp privately funded a 73-day mission to Mir by two cosmonauts later that year, which turned out to be the final expedition. Plans to refurbish the station and raise its orbit were not realized, and the Russians eventually succumbed to political pressure to decommission the Mir and concentrate its resources on the ISS. The fascinating saga of MirCorp is chronicled in Michael Potter's documentary *Orphans of Apollo*.

In "Fixer Upper", the fictional Xīn Shìjiè Company is a kind of latter-day MirCorp. But where would CEO Liǔ Diānrén have gotten an idea like trying to send the ISS to the Moon? Perhaps, somewhere in his misguided youth, Liǔ Diānrén might have read the final report of a student project from the 2003 summer session of the International Space University (ISU). Named *Metztli* after the Aztec lunar goddess, the objective of the project was to define options for robotic and human missions to the Moon using ISS capabilities where advantageous. As part of their work, the students conducted a preliminary feasibility study of sending the ISS and/or components thereof towards lunar space using an Earth-Moon cycling orbit (EMCO). The *Metztli* project is one of the most innovative to come out of an ISU session and its final report is well worth reading (http://tinyurl.com/Metztli-ISU). In 2011, *Aviation Week* reported that some engineers at NASA and Boeing were looking into the feasibility of repurposing ISS modules for potential use at an Earth-Moon Lagrangian point or in lunar orbit. Prior to the deorbiting of the ISS, Axiom intends to detach its modules as the basis of a new commercial space station. At one time, the Russians had considered detaching Nauka and Prichal with some of their older modules to create an independent space station called OPSEK, but these plans were cancelled in 2017.

Both the Americans and the Russians have experience in successfully repairing crippled space stations. Six years before dropping in on the Australian outback, the first U.S. station Skylab suffered significant damage during its launch in 1973. As a result of the mishap a micrometeoroid shield separated from the hull and tore away, taking one of two main solar panels with it and jamming the other panel so that it could not deploy. The first Skylab crew of astronauts Pete Conrad, Joe Kerwin, and Paul Weitz succeeded in installing a Sun shield over the damaged hull and releasing the stuck solar array. Equally remarkable was the courage of Russian cosmonauts Vladimir Dzhanibekov and Viktor Savinykh, who in 1985 endured almost two weeks of cold, darkness, and physical hardship to bring the Salyut 7 station back to life after a power system failure left it dead in space. A Russian film dramatizing the rescue of Salyut 7 was released in 2017.

Shortly after the launch of the first elements of the International Space Station in 1998, I remember reading an online post by *Babylon 5* creator J. Michael Straczynski in which he commented on the outward physical resemblance between his fictional station and the Zvezda service module of the ISS. As I wrote in the Preface, such linkages between science fiction and our real-life exploration of space are a big part of what appeals most to me about the genre. Historian Roger Launius has called space stations "base camps to the stars", and I cannot help but agree. Now, if only the Centauri would actually sell us jumpgate technology, then we'd really be getting somewhere.

Decrypted

Mike Petrov was getting the crap beaten out of him.

He went down quickly when the fists lashed at him, curling into a tight ball on the ground. The kicks came next, pounding his side and back as he tried to cover his face. A boot drove into his crotch, and he screamed. Then the fists returned, knuckles and the heels of hands pummelling him in the head, neck, and shoulders.

Mike didn't know it at the time, but his trouble started a long time ago, with just four little words.

Classical computer. Bits. Ones and zeros. Quantum computer. Qubits. Super-position of both ones and zeros. A classical computer performs one calculation at a time. A quantum computer can do many, simultaneously. This gives them phenomenal computational power.

"Sir, I must respectfully disagree with your opposition to Senator DeGraff's proposed Defence Against Machine Awareness legislation. Now that quantum computers are a practical reality, a ban against research into artificial intelligence is urgently needed. D-Day is merely the tip of the iceberg. Should quantum computing be applied to artificial intelligence, society would surely be on a slippery slope to –"

Mike Petrov stopped typing. *A slippery slope…to what, exactly?* He needed to think carefully about what he wanted to say—and even more importantly, how he said it.

Online discourse had civilized in the few short weeks since D-Day. Less than a month after Chandrasekara's group in Singapore announced

E. Choi, *Just Like Being There*, Science and Fiction,
https://doi.org/10.1007/978-3-030-91605-3_14

their successful demonstration of a quantum computer designed to execute Shor's algorithm, the hacker group Nì Míng leaked everything. While the world's public key encryption infrastructure did not crumble overnight, it happened fast enough for the twenty-four hour crisis news networks to declare Decryption Day.

Whatever catastrophe D-Day was to the financial sector, national security, medical privacy, and every other facet of modern society that relied upon conventional encryption, at least it seemed to have also silenced many of the world's trolls by robbing them of anonymity. Racists, homophobes, and paedophiles were having a particularly rough time.

What on Earth made them think they could shout all that swill, protected by promises of "anonymity?" Mike thought. *What...forever?*

The doorbell rang. Mike went to the front entrance and looked through the peephole, making out the distorted fisheye image of a young man in an ill-fitting uniform of the United States Postal Service.

"Greetings from the restored U.S. government!" the postman smiled, way too chipper.

"Is that a joke?" Mike growled. The USPS was enjoying something of a renaissance since D-Day. Going postal had taken on a whole new meaning.

The postman passed him an envelope. "Signature, please."

Mike closed the door and took the envelope back to his room, smiling upon recognizing his mother's name and St. Petersburg address. Inside was a greeting card, and a cheque for a non-trivial amount of money.

"Congratulations for new place," read the handwritten note. "Here is small home burning gift for wonderful son. Buy something nice."

As Mike put the cheque in his wallet, he was startled by a sudden *whap* against a window. He dashed into the living room to find a splattered egg running down the outside of the glass.

The public key encryption upon which much of our information technology infrastructure depends is based on an unproven assumption that certain mathematical problems – prime factorization and discrete logarithms, for example – are too difficult for classical computers to solve in a reasonable time. Too difficult...for classical computers.

The line of clients was stretched out into the parking lot by the time Mike got to the Northgate branch of the First State Bank of Maryland. He edged his way through the crowd, muttering apologies and occasionally flashing his employee badge in response to dirty looks. Near the entrance were four dormant automatic teller machines, their screens dark and their keypads covered with yellow tape. Once inside the building, Mike endured another round of badge flashing and apologising while navigating around impatient clients before finally arriving at the teller line.

Imani Williams, the manager of the Northgate branch, was at Mike's station. He waited until she was finished with a client before approaching.

"I am so sorry to be late," Mike said.

"Is everything all right?" Imani asked. "Did you manage to reach your landlord?"

"Yes, thank you."

"Did you call the police?"

Mike shook his head. There was really no point. Police everywhere were surprisingly busy since D-Day as mischief, from petty vandalism to occasional violence, seemed to grow in the real-world just as people were starting to get nicer online.

I guess it's an inevitable side effect, he mused. *Folks are finding out who their anonymous tormenters were, and getting even. But...but surely that's not the reason I was egged. I never did anything anonymously awful online.*

Or did I?

Imani patted Mike on the shoulder, then said in a voice loud enough for those nearby to hear, "I work for thirty-five years at this bank and this guy here busts me down to teller on my last day?"

Mike smiled as he took his seat. He did not login to the system. Like the ATMs, the computer in front of him was dark.

"Can I help the next person?"

Brick and mortar branches had been busy since D-Day, as banks and their clients abandoned online services and reverted to physical transactions. Mike would never forget that long week after D-Day, with its run on banks as frightened depositors clamoured to withdraw their savings. A near riot had broken out at Northgate, forcing Imani to call for police and extra security as tempers flared and pushing and shoving escalated to brawls. Amid the general chaos resulting from the loss of classical public key encryption, the U.S. Federal Government ordered a forty-eight hour suspension of financial transactions to, from or within the United States to give banks and businesses time to re-establish non-electronic means of commerce, including massive shipments of cash. In an emergency joint session of the U.S. Congress, the President called for the accelerated development, certification, and deployment of quantum-safe cryptographic infrastructure as an urgent national priority.

"...two-sixty, two-eighty, three hundred." Mike finished counting the polymer bills and slid them over to the client.

The young woman smiled. "Thanks so much."

"I am sorry about the wait."

"No problem." Her eyes seemed to linger on him for a moment before she turned to leave. "Take care!"

Mike rubbed his eyes. He had been on the teller line for almost six hours without so much as a restroom break. His neighbour at the next station, Dustin, handed him a bottle of water. Mouthing a silent "thank you", he gratefully took a swig before waving over the next client.

A short young man wearing darkened dumb spectacles with a hoodie and baseball cap approached. Mike's senses heightened and he shot a glance over the crowd, trying to find the security guard. He pulled himself together and calmly asked, "How can I help you today?"

The young man passed him a folded deposit slip. Forcing himself to stay composed, Mike slowly opened the paper. There were four words, neatly written in block letters:

FUCK HER, SHE'S CRAZY.

Mike read the note twice, blinking in puzzlement. Finally, he looked up and caught a glimpse of the back of the young man, calmly yet deliberately making his way to the exit. Momentarily confused, Mike couldn't decide whether to press the silent alarm or call out to the security guard. In the end, he did neither.

"Mike, are you OK?" It was Dustin. Mike held up the paper. Dustin's eyes widened and his mouth dropped in a "what the hell?" expression.

"Hey!" the man next in line yelled. "Are we getting some service here?"

Mike waved the client over with one hand, crumpling the note in the other. Its erstwhile courier was long gone.

In 1994, Peter Shor came up with algorithms for prime factorization and discrete logarithms that could run on a quantum computer. There were no quantum computers in 1994.

"I'm outta here!" Imani Williams declared. "No more of these damn Nor'easters for me. Jian-Heng and I are moving to sunny New Mexico!"

"There's an Old Mexico?" someone quipped.

Most of the Northgate staff was out at the Royal Mile pub for Imani's retirement party. Mike was sitting at a table with Dustin Miller and Natalia Lopez, nursing his third or fourth pint of Yuengling beer. It was a poor substitute for vodka.

"So, what's the story with that weird note you got this afternoon?" Dustin asked.

Mike shrugged.

"What's this all about?" Natalia asked, cradling a mug of black coffee. As the IT manager for the Northgate branch, her hours since D-Day had been particularly long. The interim solution was to physically isolate the bank's intranet from external networks. For the time being, data transfers to the outside world would be done with fax machines and trusted human "sneakernet" couriers. The pressure was on for a permanent post-quantum solution, but like everyone else Natalia was struggling with the sometimes contradictory directions from NIST, DHS, FS-ISAC, and various other acronymed bureaucracies—to say nothing of First State Maryland corporate. One day it was lattice-based cryptography, the next open avowal, and then quantum key distribution, or maybe a combination of some or all.

Dustin explained the afternoon's incident.

"Have you pissed off anyone online lately?" Natalia asked.

"Absolutely not!" said Mike. "Why do you ask such a thing?"

"You must know about this meme that's been going around since D-Day," Natalia continued. "People calling out – some would say harassing – other people they think have been anonymously rude, vulgar, racist, or bullying or whatever online. It started in Sweden. They've got a big tradition of this sort of thing over there. Way before D-Day, these Swedish vigilantes would track down crypto-masked trolls and literally knock on their doors to confront them. There was even a Swedish reality show about it at one time."

"They were going after trolls *before* D-Day?" Dustin asked.

Natalia took a sip of coffee. "It was harder then, but yeah. The obstacle was ISPs wouldn't give up their user information without a court order. After D-Day, hackers like Nì Míng could just walk in there and decrypt their account databases and DHCP logs. There could be years of records in there, thanks to Sarbanes-Oxley legislation."

"Why it is necessary to hurt people?" Mike asked, his speech a bit slurry. "It is enough to not have anonymity. People much nicer now online."

"As far as I know, nobody's actually been hurt," Natalia continued. "Some petty vandalism, sometimes face-to-face confrontations with name calling and stuff, but that's it. And no, Mike, I don't agree with you. Stripping away anonymity is necessary but not sufficient. I remember years ago there was this asinine thing where guys would yell out vulgar crap at female reporters. Most of that was caught on camera, but they didn't care. Hell, they probably *wanted* to be on camera. It didn't get better until these jerks started facing real-world consequences – getting fired, getting charged, getting banned from stuff." She took another sip of coffee. "Maybe these new vigilantes feel the same."

Imani Williams approached the table. Natalia and Dustin, and then Mike, shuffled their chairs to make room.

"Thanks for coming out," Imani said.

"We're gonna miss you, boss," Dustin said.

Imani turned to Mike. "Dustin told me about the note. I don't think there's a problem, but in any case, I'm sure we've got the creep on video if we need it."

"Blagodaryu vas." It took a moment to register, but Mike finally realized that reverting to Russian meant he was drinking too much. He called to a server. "Cheque, please!"

"Do you have cash?" the server asked.

"No cash. Credit card."

The server muttered something, then excused himself to get the manual credit card imprinter from behind the bar.

Classical public key cryptography has one more shortcoming. Quantum computers are able to break it retroactively. Years of information would be revealed.

It was well past midnight by the time Mike stepped off the bus. He staggered a few steps to the shelter and leaned against it for a moment. A light rain was falling, which he found an oddly soothing balm for his inebriation.

"Are you OK, sir?" the driver called down.

Mike nodded and waved him off. The door closed and the bus pulled away, quickly disappearing into distance and darkness. Taking a few deep breaths, Mike staggered in the general direction of home. It did not take long for him to wander from the lights of the street.

There were two of them. Even in his intoxicated state, Mike knew he was being followed. His pursuers made no attempt at stealth. They called out to him.

"Mr. Petrov!"

Mike quickened his pace.

"We just want to talk to you, Mr. Petrov."

Mike started to run, but after only a few feet he tripped on a bit of upraised pavement and fell to the ground.

They were upon him immediately. Hands grabbed his shirt, pulling him onto his feet.

"You're Petrov, right?"

They were two young men of medium build. One wore a sweatshirt, the other a hoodie. Mike could not tell if the latter was the same guy who had been at the bank that afternoon.

Sweatshirt guy grabbed Mike by the chin. "It's him – mikhailp36."

It took a moment for Mike to recognize his old online handle. Six or seven years old, at least. The mikhailp36 handle probably dated back to his arrival in the United States.

"Ya ne ponimayu."

"I have no idea what you're saying," sweatshirt guy said, his breath also smelling of alcohol. "So you just listen close to these four little words: 'Fuck her, she's crazy.'"

Mike stared at his accusers with blank incomprehension.

"Your post." Hoodie guy was speaking now. "That was you, seven years ago, responding to a post by a woman whose daughter was undergoing naturopathic treatments instead of chemo for leukaemia. You…yours was the index case, the post zero that started the crap storm of online abuse against her."

Mike struggled to remember. He had just arrived in the United States, trying to read online forums as a way of helping him learn English. At the time, he would have made pretty regular use of translation tools to help him along, both to read and on occasion to write.

"A mother," hoodie guy shouted, "trying to take care of her sick child!"

Ona boleyet…she is sick. Perhaps the translation tool confused it with ona bol'naya—she is crazy.

Mat'…mother. Mat' yeye…care for her, mother her.

Tvoyu mat'. Fuck.

"Oshibka!" Mike exclaimed. "Big mistake!"

"Unbelievable," sweatshirt guy sneered. "Absolutely no remorse. Big mistake my ass."

That was when Mike tried to take a swing at him. Maybe it was the alcohol, maybe it was fear, maybe he just wanted the guy to shut up. Something snapped, and he tried to throw a punch. The fist met only air.

His opponent was not so inept and sufficiently less intoxicated. The blows quickly put Mike on the ground. Sweatshirt guy then started kicking him, a boot landing squarely into his crotch. Mike screamed as sweatshirt guy continued to pummel his head, neck, and shoulders.

"Jesus Christ! Anton, what the hell are you doing?" Hoodie guy was trying to pull his companion off the helpless Mike.

"Gotta show these bastards –"

"It's not like this!"

"Back off, Noah!" Anton raised a fist. "I swear to God, I will tear –"

"*What the hell's going on here?!?*"

All three men turned to the sound of the new voice. A young woman was standing about thirty metres away. She started walking towards them, slowly but without fear, and they could see she was wearing smart glasses. As she got closer, they could discern the faint green electronic intention light on the rim that was the tell-tale of AI-ware.

"Let's go," Noah said to Anton. "We're done here."

She waited for the two men to have gone a reasonable distance before approaching Mike, who was still sprawled on the ground. Gently, she touched his shoulder.

"Are you all right?" She started helping him get to a sitting position.

Mike reached out, but his flailing arm accidentally smacked the side of her head and knocked her AI-ware to the ground.

"Izvinite. Sorry, so sorry."

"It's all right," she said. "They're not even real. I can't afford real digi-spectacles." She pulled an ancient flip-style cell phone from her purse and dialled 911. "Yes, hello. I'm here with a man who's been assaulted." She looked around. "Um, we're on Bel Pre Road, near Aquarius Park…Yeah, he's hurt, but he's conscious and sitting up…Right. Yes, of course. Of course I'll stay with him."

The world's IT infrastructure relies upon cryptography, physical security, and trust. Trust is the hardest.

A bouquet of flowers, courtesy of his Northgate colleagues, bathed in sunlight streaming through the clean living room windows. Beside it was a wish-you-well card from somewhere in New Mexico.

At his computer, Mike began to type. "To the young woman who helped me last Friday night. You said you want to be anonymous, but what you did for me should not be. Thank you. Spasibo. If you wish, please contact me. I have a small gift. Mikhail 'Mike' Petrov (mikhailp36), Silver Spring, Maryland."

Next to him on the desk, linked to the computer, was a pair of brand new digi-spectacles.

Afterword

The Decryption Day catastrophe portrayed in the story is a very real threat. Much of our current information technology (IT) infrastructure depends on public key encryption that is based on an assumption that certain mathematical problems—for example, factoring the product of two prime numbers back into the original primes—are too difficult for classical computers to solve in a meaningful time. For example, it is estimated that it would take a century for a classical computer to break 1024-bit RSA.[1] But the assumption of mathematical complexity behind problems such as prime factorization, discrete logarithms, or elliptic curve schemes is actually not proven. A sudden or unexpected algorithmic innovation would immediately comprise many modern security systems.

In 1994, the American mathematician Peter Shor devised algorithms that would enable quantum computers to break cryptographic systems based on prime factorization and discrete logarithms. A classical computer use bits that represent either a one or a zero and can perform one calculation at a time. In contrast, a quantum computer uses "qubits" that are a superposition of both ones and zeros, giving them phenomenal computational power with the ability to perform many calculations simultaneously. There were of course no quantum computers in 1994, however, the prevailing scientific consensus is that practical and widespread quantum computing capabilities will likely be available within the next 10–25 years.[2] When that happens, most current encryption methods will be rendered useless.

Another problem is that most present day cryptographic systems are not "future-proof". Many organizations such as health care institutions or the military need to secure sensitive data over the long-term, for example, for twenty years or more. An unexpected algorithmic innovation or the advent of quantum computers would not only compromise contemporary data but would also make years of historical records vulnerable.

With regard to the specific threat of the quantum computer, one may ask if it is really something that needs to be a concern at the present time. To answer this question, it may help to express the problem in the language of mathematics. Let x be the number of years that cryptography must remain unbroken (i.e. how long it is needed to protect health information, or national security information, or trade secrets). Let y be the number of years it will likely take to replace the current system with one that is quantum-safe or not based on unproven assumptions of mathematical complexity. Finally, let z be the number of years it will take to break current encryption tools using quantum computers or otherwise. If $x + y > z$, then there is a problem right now and immediate action must be taken.

There are two possible solutions. The first are "post-quantum" or "quantum resistant" techniques, which are new codes and protocols based on

---

[1] Corker, D., Ellsmore, P., Abdullah, F., Howlett, I. "Commercial Prospects for Quantum Information Processing", Quantum Information Processing Interdisciplinary Research Collaboration, 1 December 2005.

[2] Schlosshauer, M., Kofler, J., Zeilinger, A. "A Snapshot of Foundational Attitudes Toward Quantum Mechanics", Stud. Hist. Phil. Mod. Phys. 44, 222–230 (2013).

"NP complete problems" for which there are currently no known quantum algorithms capable of breaking them. While post-quantum methods would not require a lot of new hardware and could be deployed quickly on existing IT infrastructures, the inherent vulnerability of being again based on unproven mathematical assumptions would remain.

Quantum cryptography is the other possible solution. It has been argued that quantum mechanics is possibly the most thoroughly tested model of the Universe that has been devised by science. For data security over the long-term, only systems based on quantum cryptography would be secure against both classical and quantum computers. This is due to the fact that quantum cryptography relies upon fundamental laws of physics rather than unproven assumptions of mathematical complexity, and hence will never be threatened by innovative algorithms or increased computational power.

An unconditionally secure or "perfect" cryptosystem is one that cannot be compromised even with unlimited time and computational power. The standard example of a perfect cryptosystem is the Vernam cipher or one-time pad. In this technique, each bit or character of a message is encrypted by combining it with a corresponding bit or character from a random secret key (the pad) using modular addition. If each key is used only once (one-time), is truly random and unpredictable and is at least as long as the message itself, then the resulting encryption will be impossible to break.

There are a number of practical problems with the Vernam cipher. First, classical computers do not actually generate truly random numbers (it would be more correct to describe them as "pseudo-random"). Second, the security of the Vernam cipher is only as good as the security of the key exchange itself. Even today, secret keys for many financial, military, and diplomatic applications are often delivered by "sneakernet" using trusted human couriers. In addition to the risk of compromise during transit (for example, a pickpocket swiping, copying, and replacing the pad), there is no guarantee the couriers themselves can be trusted. Anyone who manages to copy or steal the key would be able to decrypt messages effortlessly, thereby defeating the purpose of encryption.

Both of these problems could be solved by quantum mechanics. Contrary to Newtonian or classical physics, quantum physics is fundamentally non-deterministic by nature. Quantum random number generators exploit the intrinsic randomness of phenomena such as radioactive decay or photonic transmission and reflection to generate true random numbers. Another technique called quantum key distribution (QKD) could solve the key exchange security problem. QKD establishes highly secure keys between distant parties by using single photons (or similarly suitable quantum objects) to transmit each bit of the key. Since single photons behave according the laws of quantum mechanics they cannot be tapped, copied, or directly measured without leaving tell-tale signs of manipulation.

Another interesting property of quantum key distribution is "universal composability", which means that a portion of a key generated by QKD could be used to authenticate the messages in a subsequent round of QKD with no decrease in security. Even if the original authentication keys are compromised after the first QKD exchange, the subsequent keys generated by QKD would remain secure.[3] In contrast, classical public key exchange schemes do not have this feature. An eavesdropper who logs all communications and subsequently breaks the first key would be able to read all future communications. But with QKD, keys for new sessions are completely independent of all prior keys and messages.

It is indeed ironic that quantum mechanics represents both the greatest threat to classical cryptography as well as the best means of securing it. Within the next two decades, it is expected that practical quantum computers will be able to routinely and quickly solve the mathematical problems upon which most current encryption methods are based, rendering them useless. To prevent the catastrophe of Decryption Day, the time for action is now.

---

[3] Stebila, D., Mosca, M., Lütkenhaus, N., "The Case for Quantum Key Distribution", arXiv:0902. 2839v2 [quant-ph] 2 Dec 2009.

A Sky and a Heaven

"There is a sky, and there is a heaven. The sky is a matter of height. Heaven is a matter of depth."—Shimon Peres

Bergen-Belsen
March 1944

Joachim Joseph was nine-years-old when the Germans invaded the Netherlands. He remembered his father looking out the window at the columns of Nazi armour outside their home. His father was crying.

They were sent to the death camp at Bergen-Belsen. Joseph was separated from his brother and his parents, each assigned to a different barrack. The poorly-constructed wooden buildings were dank and filthy. Alone and hungry, Joseph lay on the straw mattress, curled in a foetal position as if trying to squeeze warmth out of his thin prison garb.

"Hello, young man."

Joseph looked up and saw the gaunt, bearded, and bespectacled face of Rabbi Simon Dasberg. One of his eyes was swollen, and there was dried blood around an ear. Joseph had heard that Rabbi Dasberg was often beaten for saying prayers outside the crematorium and refusing to shave his beard.

"What is your name?" Rabbi Dasberg asked.

"Joachim Joseph."

"And how old are you, Joachim?"

"Twelve."

"When is your birthday?"

© The Author, under license to Springer Nature Switzerland AG 2022
E. Choi, *Just Like Being There*, Science and Fiction,
https://doi.org/10.1007/978-3-030-91605-3_15

"March 31st."

Rabbi Dasberg smiled. "I thought you looked Bar Mitzvah age." He lowered his voice conspiratorially. "Would you like a real Bar Mitzvah ceremony?"

As Joseph watched in amazement, Rabbi Dasberg opened a little box and un-wrapped a cloth to reveal a four-inch Torah scroll. "I can teach you to read from this, if you are willing."

For the next few weeks, Rabbi Dasberg woke Joseph early every morning, and together they studied the text of the tiny scroll. From their bunks, some of the other men looked on. A few even smiled at the defiant little conspiracy that was taking place.

"Well, the time has come," Rabbi Dasberg said one morning, a few days before Joseph's thirteenth birthday.

The men peered outside, looking for the guards, and then covered the windows with blankets. Four candles were lit, their faint light flickering in the pre-dawn darkness. Rabbi Dasberg placed a small towel on a table, put down the Torah scroll, and unrolled it.

"Stand up, Bar Mitzvah boy," Rabbi Dasberg said softly.

With tears in his eyes, Joachim Joseph approached the table and took hold of the scroll. Rabbi Dasberg and the other men huddled around the boy. Joseph opened his mouth, and poised himself to read.

Someone knocked on the door.

Near Baghdad, Iraq
June 7, 1981

Captain Ilan Ramon held the stick of his Israeli Air Force F-16 fighter jet with intense focus, his other hand resting on the throttles. The head-up display in front of him indicated an airspeed of 450 knots and accelerating, and an altitude of 102 feet above ground level. Outside his canopy, he could see the other seven fighters of the attack squadron in tight formation. Ramon was "flying the tail", the last plane of the group.

They had taken off from Etzion Airbase over an hour and a half ago, and were now just twelve miles from their target—the Osirak nuclear reactor southeast of Baghdad. Ramon selected full afterburner and pulled back on the stick, following the planes ahead in a climb up to 6,000 feet. They were now visible to Iraqi radar. At 6,000 feet they pitched over into a dive, aimed at the containment dome of the reactor complex.

The planes ahead released their bombs in pairs at thirty second intervals. Ramon stood up the throttle and flicked the master arm switch. Smoke, explosions, and anti-aircraft fire were already obscuring the target. He pressed the weapon release button, feeling a kick in his seat as the bombs fell away.

Ramon pulled back on the stick and advanced the throttle, bringing his plane to a banking climb away from the smouldering reactor. He looked over his shoulder, and at the periphery of his vision spotted the serpentine plumes of two surface-to-air missiles.

There was an explosion.

The control panel lit up. The engine flamed out. Stick and throttle no longer responded.

Ramon grasped the ejection handle and pulled. The canopy flew away with a bang, and then a small rocket motor blasted his seat clear of the disintegrating plane.

Ilan Ramon fell into the sky, and was thrown into darkness.

National Air and Space Museum
Washington, DC
Spring 1995

Five-year-old Dean Issacharoff looked up at the sky in wonder.

Suspended from the ceiling, historic aircraft floated in silent majesty. Dean could see the bright orange Bell X-1 that broke the sound barrier, and the silvery grey *Spirit of St. Louis* flown from New York to Paris by the American aviation pioneer (and anti-Semite) Charles Lindberg.

Jeremy Issacharoff, a political counsellor at the Embassy of Israel in Washington, took his son's hand. "Hey Dean, let's see the space stuff over there."

They ambled past the Apollo 11 command module and the Skylab orbital workshop, and soon found themselves before a model of the Space Shuttle.

"Check it out."

"Cool!" Dean smiled as he surveyed the Shuttle model, the white-and-black orbiter straddling the large orange external tank, flanked by a pair of white solid rocket boosters.

"Look at this." Issacharoff gestured at a display. "Here are some of the astronauts who have flown on the Shuttle. They're from many countries." He pointed. "What's that flag there? The one with the red maple leaf?"

"Canada."

"That's right. Mr. Garneau is from Canada. And the blue, white and red flag? Where is Mr. Baudry from?"

"France."

"And this green flag with the sword. Do you know what country that is?"

The boy shook his head.

"It's Saudi Arabia. A prince from the Saudi royal family once flew on the Shuttle."

Dean furrowed his brow. "Papa," he said. "Why are there no astronauts from Israel?"

Directorate for Defence Research and Development (Maf'at)
Israel Ministry of Defence, Tel Aviv
February 1996

"*Is this Yael Dahan?*" asked a gruff sounding voice on the phone.

"Yes," the research engineer replied. "Who is this?"

"*I am Aby Har-Even. I am director general of the Israel Space Agency.*"

"Hello." Dahan gripped the phone tighter, a frisson of excitement sweeping over her.

"*Let me get to the point. Would you like to be Israel's first astronaut?*"

Her jaw dropped. She stammered something nonsensical.

"*What is it?*" Har-Even barked. "*Is there a problem?*"

"N-no, no problem."

"*You did express an interest, did you not? You did submit an application, did you not?*"

"Yes, of course." Dahan composed herself. "It's just that...well, I'm a little bit surprised. What I mean is, I was expecting some kind of formal selection process, a competition –"

"*Listen to me,*" Har-Even said. "*We do not have the time or money for an extended search. Peres and Clinton announced the astronaut opportunity last year, and now we need to put forward a candidate immediately to make the start of this year's NASA astronaut class for mission specialists.*"

"I...see."

"So, Yael Dahan," Aby Har-Even said. "Are you in?"

Department of Geophysics and Planetary Sciences
Tel Aviv University
Summer 1996

Professor Joachim Joseph walked into the conference room for the weekly meeting of his research group. He sat down at the head of the table.

"Yoya, you are smiling," said the summer student Mustafa Asfur, calling Joseph by his nickname. "You have good news?"

"Yoya is always smiling," said Meir Moalem, a master's candidate.

"Indeed, I have good news," Joseph said. "Our MEIDEX payload has been selected to fly on the Space Shuttle mission with the Israeli astronaut." The objective of MEIDEX, the Mediterranean Israeli Dust Experiment, was to study how dust particles blown off the Sahara Desert affect climate and weather across North Africa, the Mediterranean, and the Atlantic Ocean. "Launch is scheduled for early 2000."

"This is so exciting!" exclaimed Asfur.

Joseph's smile broadened. "Yes. This will be an amazing scientific adventure."

Ramat Aviv Gimmel, Israel
January 2002

"Thanks for the ride," Aby Har-Even said in the grumpy old man voice that Yael Dahan had grown to adore.

"No problem," said Dahan's husband, Omer Bitten. "We're glad you could join us for Shabbat dinner with Yoya."

"It's the least we can do," she added. "After all, I owe you for *my* ride."

Har-Even harrumphed. "Yes, well, you can be sure there will be discussions about that."

"Is there a problem?"

"It's these delays, Yael," Har-Even sighed. "You were supposed to have flown almost two years ago. Frankly, we are running out of money to keep you in the programme. And if that wasn't enough, I get letters from Haredis who object to a woman representing the Jews in space."

"Well, they don't have to sit next to me on the Shuttle if they don't want to."

Noa, her fifteen-year-old daughter, giggled.

When they arrived at the Joseph home, the professor and his wife Stella greeted them. They all went into the dining room.

Stella lit the Shabbat candles, covered her eyes, and recited. "Baruch Atah adonai Elhenu melech haolam asher kideshanu bemitzvotav vetzivanu lehadlik ner shel Shabbat kodesh."

As they ate, Dahan looked up and saw a little wooden ark sitting on a shelf.

"Yoya, what is that?"

Joseph got up from the table and took the ark down from the shelf. He opened it, revealing a tiny four-inch Torah scroll inside.

"It's beautiful," Noa said.

Dahan looked at Joseph. "There must be a story."

Bergen-Belsen
March 1944

Someone knocked on the door.

Everyone froze. Rabbi Dasberg went to the door and slowly opened it.

Joseph's eyes widened.

"Eema!" he cried, throwing his arms around his mother's emaciated frame.

"It is dangerous for you to be here," Rabbi Dasberg said quietly. "Women are not allowed in the men's barracks."

The boy held his mother tight. Gently, she finally pushed him away and touched his cheek. "I will stand outside the window and listen."

Joseph nodded, wiping his eyes. Rabbi Dasberg led him back inside to continue the ceremony. "Baruch Atah..." he began, reciting the blessings and reading from the Torah. Rabbi Dasberg cupped his hands on Joseph's head and blessed him. The men rose from their bunks to congratulate the boy. One gave him a sliver of chocolate, barely larger than a thumbnail, wrapped in a small piece of paper.

Rabbi Dasberg took Joseph aside, and to the boy's surprise, pressed the little Torah into his hands. "I want you to take this."

Joseph gasped. "How can I take a Torah?"

"Listen to me," Rabbi Dasberg said. "You must take this, because I will not leave here alive. You must promise that when you get out, you will tell the story of what happened here. Do you promise?"

Joseph looked at Rabbi Dasberg. "I promise."

Ramat Aviv Gimmel
January 2002

A solemn silence fell on the dinner table. Dahan's head was bowed, and the eyes of her daughter and husband were moist.

"Yoya," Dahan said at last, "I must tell you something. My mother and father were also 'graduates' of Bergen-Belsen. I am an only child. My parents waited quite late in life before having me. They were almost forty when I was born. I always got the sense they were hesitant to bring a child into the world, a world that seemed for so long to be without goodness and hope."

Dahan looked squarely at Har-Even. "I will be going back to Houston soon, to continue training for my flight. I look forward to operating the MEIDEX experiment and getting good science. But more than that, more than that..." She turned to Joseph. "Yoya, I wish to ask you – may I please have the honour to bring your Torah with me to space?"

The man who was the boy at Bergen-Belsen said yes.

Dahan gazed at Har-Even. He closed his eyes, and nodded.

Mission Control Houston
NASA Johnson Space Centre, Building 30
January 16, 2003

Smiles, waves, and the occasional call out of "hello" and "good morning" greeted Ilan Ramon as he entered Mission Control. He took his place at the Flight Director's console, resting his cane against a filing cabinet. This would be his eighth Space Shuttle mission and his first as the Lead Flight Director. The screen at the front of the room displayed a live feed from the Kennedy Space Centre showing the Shuttle *Columbia* sitting on the launch pad.

The countdown clock was stopped, holding at T-minus 9 minutes.

Ramon adjusted his headset and keyed the mike. "Good morning, STS-107 flight controllers. Please give me a go/no-go for launch."

"FDO?" "Go."

"EECOM?" "Go, Flight."

"INCO?" "We are go."

"CAPCOM?" "We're go, Flight."

After polling the rest of the flight controllers, Ramon switched to an external loop. "Launch Control, this is Houston. You are 'go' for launch."

Kennedy Space Centre, Florida
10:39 am EST, January 16, 2003

Yael Dahan's family gathered with those of the other *Columbia* astronauts to watch the lift-off from the roof of the Launch Control Centre. Beside Noa and Omer were Yael's parents Eitan and Michal, survivors of Bergen-Belsen who had endured the depths of hell and were now about to watch their daughter soar into the heavens. Noa saw her stepfather's parents Moshe and Rivka and his sister Neta talking quietly to Stephanie Wilson, her mother's friend and fellow graduate from the NASA Astronaut Class of 1996.

"*T-minus 31 seconds, and we have a 'go' for auto sequence start,*" announced the voice from the loudspeaker. "*Columbia's on-board computers have primary control of all the vehicle's critical functions.*"

Holding Omer's hand, Noa walked closer to the railing.

"*T-minus ten...nine...eight...seven, we have a 'go' for main engine start...*"

From the distant launch pad, a bright flash appeared at the base of *Columbia*, obscured a fraction of a second later by a rising column of steam.

"...three...two...one...we have booster ignition and lift-off of the Space Shuttle Columbia with a multitude of national and international space research experiments."

A second dawn lit the horizon. Cheers erupted from the crowd as Columbia roared off the launch pad and started rolling onto its back, angling upwards in a north-easterly direction over the Atlantic Ocean.

Noa squealed and jumped like a child, covering her mouth with her hands. "L'hitraot, eema! Kol tuv, eema!"

NASA Johnson Space Centre, Building 13
Houston, Texas
January 17, 2003

Engineers stared at a screen on the far wall, mesmerized by a grainy film that was being played in a repeating loop. Over and over, the film showed an object breaking away from *Columbia*'s external tank and smashing into its left wing, bursting into a shower of particles.

Ilan Ramon recognized Rodney Rocha, the division chief for structural engineering, and approached him. "What do we have here?"

"Something, probably a piece of insulating foam, came off the external tank about 82 seconds after launch and hit the left wing of the vehicle," Rocha said.

"Is there a danger?" Ramon asked immediately.

"We don't know," Rocha replied. "It depends on the size of the foam, the speed and angle of impact, and where it hit. Based on this footage, our preliminary estimate for the foam size is between 21 to 27 inches long and 12 to 18 inches wide, with an estimated impact speed somewhere between 625 and 840 feet per second relative to the vehicle."

"And the impact location?" asked engineering manager Joyce Seriale-Grush.

Rocha picked up a model of the Shuttle and tapped his finger on the leading edge of the left wing. "Somewhere here, probably on the reinforced carbon-carbon panels, or maybe on the thermal protection tiles just below."

"That is a lot of uncertainties," Ramon said.

"The uncertainties are due to the low frame rate and poor resolution of the camera," Rocha explained. "These views were taken by a film camera seventeen miles away."

"What does your team need to resolve the uncertainties?" Ramon asked.

"Better images," Rocha said, without hesitation.

"Can the astronauts see that area, by looking out the windows?"

"No. The open payload bay door blocks the view."

"You know," Seriale-Grush said, "back in the early days of the Shuttle programme, they sometimes got spy satellite images of the Shuttle in orbit to see if the thermal protection tiles were all right."

Rocha's eyes widened. "Can we do that? Can we petition for outside agency assistance?"

The question was met with silence. Talk about classified capabilities was off-limits. Rocha rephrased his request. "We can't tell whether there's damage to *Columbia*, and we need pictures of the vehicle in orbit."

"Let me take an action to bring this up with the MMT," Ramon said, referring to the Mission Management Team.

Payload Operations Control Centre
NASA Goddard Space Flight Centre
Greenbelt, Maryland
January 19, 2003

"Wonderful!"

With each new set of MEIDEX data coming down from *Columbia*, Joachim Joseph called out and pointed at the screen with delight. The instrument had captured images of dust plumes off the coasts of Nigeria, Mauritania, and Mali. A colour scale showed the dust concentration from green for thinnest to red for heaviest.

"Look at this, Yoya." Scott Janz, his NASA co-investigator, pointed at a greyscale contour plot. "I'd say we've captured ourselves some ELVES," he said, referring to electromagnetic perturbations in the upper atmosphere.

The new message flag appeared. Joseph opened the email and double-clicked on the attachment. A low-resolution video of Yael Dahan with Commander Ben Hernandez floating in the middeck of *Columbia* began to run.

"*Good morning, Yoya,*" Hernandez said. "*From the crew of* Columbia, *congratulations on the successful activation of MEIDEX. We hope you're getting good data. And we know you'll be heading back to Israel later today, so we'd like to offer this little poem.*"

Hernandez passed the microphone to Dahan, who began reading from a piece of paper:

"*Yoya Joseph, our friend on the ground,*

"*We send him our data as we go round and round,*

"*His warm friendly eyes make us all realize,*

"*What a wonderful friend we have found.*"

Dahan and Hernandez waved at the camera. "*N'see'a tova, Yoya! From the crew of* Columbia, *best wishes for a safe trip home.*"

NASA Johnson Space Centre, Building 13
January 19, 2003

Ilan Ramon hated PowerPoint.

Over the course of his twelve year career at JSC, he watched in dismay as the use of PowerPoint presentations instead of technical reports became endemic. PowerPoint was fine for marketing hucksters, but when complex engineering analyses were condensed to fit into bullets on standardized templates, crucial information was inevitably lost.

A young engineer was presenting the results of the impact analysis his team had performed using a computer model called Crater. On one slide, the vague words "significant" and "significantly" were used five times without quantification. The slide also had "cubic inches" written inconsistently: "3cu. In", "1920cu in" and "3 cu in". Such inconsistencies and lack of precision were antithetical to Ramon's military background, where a misplaced decimal or mistaken unit of measurement could have deadly consequences.

"We've seen these foam impacts on past missions, and it's never been a problem." The speaker was Peter Delaney, a senior engineer with thinning grey hair and Coke-bottle glasses who was regarded by management as an expert on the Shuttle's thermal protection system. "As the Crater analysis has confirmed, there is no threat to *Columbia*. We are dealing with a post-flight maintenance issue, not a safety-of-flight issue."

Ramon, Joyce Seriale-Grush, and Rodney Rocha looked at each other.

"Wasn't there a slide that said Crater predicted damage exceeding the thickness of the thermal tile?" Seriale-Grush asked, her voice a mixture of confusion and alarm.

"The Crater model is very conservative," Delaney explained. "It doesn't take into account that the thermal tiles are denser at the bottom, so it's always predicted greater damage than what actually occurs. In any case, as I've said, the impact on *Columbia* is in-family with similar events on previous flights."

"I'm sorry, but that's a lousy rationale," Seriale-Grush said. "Things aren't supposed to be hitting the Shuttle at all. Just because it's happened before without consequences doesn't mean there's no danger."

"Look," Delaney hissed, "I was there when they invented those tiles and I've got a dozen tile-related patents to my name, so I know what I'm talking about."

"Nobody is questioning your expertise, Peter," said Ramon, trying to moderate, "but Joyce raises an important point."

Rocha spoke up. "Well, *I'm* going to question. Peter's expertise is in the tiles, but the film analysis indicates it was more likely the foam hit the RCC panels," he said, referring to the reinforced carbon-carbon leading edge of the Shuttle's wing.

"The RCC is even more resilient to damage than the tiles," Delaney said. "If that's what you saw, it's the best thing that could've happened."

"Peter, I grant your expertise in what you know and your experience with the tiles," Rocha said, his voice starting to rise. "But you are not a structural dynamicist. You are not a stress expert. You are not an image specialist for enhancing pictures. You are none of these things. So how can you know?"

Delaney started to respond, but Ramon held up a hand. "That is the fundamental issue here. None of us actually know." He pointed at the screen. "Those last two bullets on the slide: 'Flight condition is significantly outside of test database / Volume of foam is 1920cu vs 3 cu in for test.' *That* is the key issue. The estimated size of the debris that hit *Columbia* is *640 times* greater than the data used to calibrate the Crater model on which we are trying to base our damage assessment. This entire exercise is invalid. We will need to do the analysis all over again when we get the images of the vehicle in orbit."

Everyone turned to look at Ramon.

"What images?" Delaney asked.

"Joyce and Rodney and I talked about it, petitioning outside agency assistance for imagery from national technical means." Ramon was using the politically correct words for spy satellites and high-resolution military telescopes. "I emailed the MMT, emailed Maria Garabedian, after we talked. Are you saying nobody's heard anything?"

Ramon called Garabedian, the chair of the Mission Management Team, as soon as he got back to his office. It went to her voicemail. He put down the phone, thought for a moment, and then started writing an email asking for on-orbit imagery to be put on the agenda for the MMT meeting first thing Monday morning.

Aboard Columbia
Flight Day 5
January 20, 2003

Yael Dahan gazed down upon the Holy Land from the heavens.

From orbit, Israel looked exactly as it did on a map. It had come into view and would be gone again in less than a minute, testament both to the speed of her spacecraft and the size of her homeland. She pressed her hand against the window, and just like that, all of it—from the Mediterranean to the Sea of Galilee to the Dead Sea—six-and-a-half million people and three thousand years of history, was gone. It was a humbling revelation.

She recited the Shema prayer. "Shema Yisra'eil Adonai Eloheinu Adonai echad…"

NASA Johnson Space Centre, Building 30
January 20, 2003

Maria Garabedian convened the Mission Management Team meeting at precisely 8:00am. Among the attendees were Peter Delaney and Mission Evaluation Room manager Don McCormack. A number of others participated by teleconference including Ron Dittemore, the Space Shuttle programme manager, and Wayne Hale, a manager at the Kennedy Space Centre. Ilan Ramon, Joyce Seriale-Grush, and Rodney Rocha sat at the back of the room.

"Good morning," Garabedian said. "Flight ops don't respect statutory holidays, so I appreciate everyone being here."

Ramon watched in amazement as Garabedian prosecuted the meeting with brutal efficiency. She seemed to revel in briskness, speaking in a stilted syntax of short-hands and acronyms like some NASA auctioneer. A/C problem in the SPACEHAB? Check. Leak in the WLCS? Check. Ku-band antenna glitch? Check.

Discussion of the foam strike was left to the end.

"We received data from Rodney Rocha's team on the potential range of sizes and impact angles and where it might have hit, and Peter Delaney's team has done an analysis," Don McCormack reported. "While there is potential for tile damage, our thermal analysis does not indicate burn-through, only localized heating."

"No damage and localized heating means tile replacement," Garabedian summarized.

"Now, the foam might have hit the RCC," McCormack continued, "but there shouldn't be anything more than coating damage."

"So, I'm hearing from Don and Peter there will be no burn-throughs, so no safety-of-flight issues." Garabedian looked at the wall clock. "All right, it's 8:30. Thanks, everyone."

Rodney Rocha turned to Ramon in disbelief. "This is ridiculous. Is that all they're going to say about it? And what about the imagery request?"

The crowd filed out of the room. Ramon got up and followed Garabedian to the elevator.

"Excuse me, Maria," Ramon said. "May I please have a word with you?"

"What do you want?"

"I had emailed you about on-orbit imagery. Why was it not discussed?"

"Is there a mandatory requirement for imagery?" Garabedian asked.

"Mandatory?" Ramon blinked. "I'm not sure what you mean."

"Is there a mandatory requirement for imagery? Because I didn't see anything in that analysis that would indicate a mandatory need."

Ramon thought he understood. "Maria, you're putting my team in an impossible situation. We need images to be able to accurately assess potential damage, but you're telling me that we need to prove the damage is bad enough to justify taking images."

The elevator doors opened and Garabedian stepped inside. "I can't help you until you show me a mandatory requirement," she said as the doors closed.

Ramon's mind churned as he drove back to his office in Building 4. He sat at his desk for a long while in contemplation, and then he made a decision. It took some time to find the right contact, but when he did, he started writing an email.

Aboard Columbia
Flight Day 6
January 21, 2003

"*Jerusalem, this is Houston,*" said the voice of the NASA public affairs officer. "*Please call* Columbia *for a voice check.*"

Yael Dahan, along with commander Ben Hernandez, flight engineer Kalpana Chawla and mission specialist David Brown, floated together in the middeck of *Columbia*. They stared at a small camera mounted on the bulkhead.

"*Hello*, Columbia," called the voice of Israeli Prime Minister Ariel Sharon. "*I am here with technology minister Limor Livnat, and you should know that our conversation is being broadcast live across the country.*"

"Thank you, Mr. Prime Minister." Hernandez spoke into the microphone. "It is an honour to be speaking with you and Minister Livnat today."

"*The honour is ours, Commander Hernandez,*" Sharon continued. "*I wish you continued success in your mission, and upon your return I would like to invite your crew to visit us in Israel. I can assure you that you will find yourselves among friends.*"

"Thank you, Mr. Prime Minister. We appreciate the invitation. And now, I would like to turn it over to Yael Dahan." Hernandez passed her the microphone.

"*Yael Dahan,*" Sharon said, "*on behalf of our entire nation, I cannot tell you enough how wonderful it is to know that a daughter of Israel is up there among the stars.*"

"Thank you, Mr. Prime Minister."

"*How does Israel look from space?*" asked Livnat.

"It is very beautiful, yet also small and fragile, and peaceful," Dahan replied. "And that is my greatest hope, for peace. Peace in our homeland, peace with our neighbours, peace throughout the world."

"*I understand you brought some special artefacts with you to space,*" Sharon said.

"Yes, Mr. Prime Minister." Dahan turned to her colleagues. "Begging the pardon of my crewmates, I will now say a few things in Hebrew.

"Mr. Prime Minister, I have one particular artefact for which I received special permission to bring. This artefact is very special."

Yael Dahan held up a tiny Torah scroll, bringing it into view of the camera.

"We are usually not permitted to show these, but today I am filled with excitement while I hold this in my hand. This little Torah scroll you see here, almost sixty years ago when he was only a boy at Bergen-Belsen, our mission scientist Joachim Joseph – Yoya – received it from the Rabbi of Amsterdam who was preparing him for his Bar Mitzvah. I am thankful to Yoya and very touched to have it here with me in space."

Ramat Aviv Gimmel, Israel

"*This little Sefer Torah,*" Yael Dahan said on the television, "*shows the ability of the Jewish people to survive anything, even the darkest of times, and to always look forward with hope and faith for the future.*"

As Joachim Joseph watched, the little Torah slipped from her hand, and for a moment, floated free. Rabbi Dasberg's Torah. In the heavens. Free.

"Yoya."

Joseph turned. Sunlight through the window caught the grey in Stella's hair, giving the appearance of a halo. She was holding a camera.

"Go over there." She waved her free hand. "By the TV."

Slowly, Joseph got up from the chair and went to stand by the television, resting his hand on the set. The little Torah was still there. And Stella took the picture.

NASA Johnson Space Centre
January 21, 2003

Being called to Maria Garabedian's office in Building 1 was, in some ways, like being summoned to the principal's office. Ilan Ramon was in trouble.

"Why did you email the NIMA POC about on-orbit imagery?" she asked, referring to the National Imagery and Mapping Agency. "Those requests are supposed to go through proper channels. Through me."

Ramon sighed. As a former fighter pilot, nobody needed to tell him the importance of following the chain of command. But even he found the management culture at JSC to be closed and insular compared to the Israeli Air Force. He believed the safety of a crew under his watch was in danger, and he had to act.

"We did go through you," Ramon said quietly. "Twice. There was no response."

"Because neither you nor anyone on your team ever articulated a mandatory requirement for imagery," Garabedian said.

"We thought the rationale was obvious." He waved a hand. "We have all these safety posters everywhere saying things like, 'If it isn't safe, say so.' Well, Maria, we really think it's that serious."

Garabedian's eyes narrowed. "My husband is an astronaut. Don't you ever imply that I don't care about safety."

"Yes, of course. I apologize."

"Listen, Ilan," Garabedian said. "This is a busy flight, with all those science payloads. There is no mandatory requirement for imagery. And you know, even if there really is damage, there's nothing we can do about it."

"Are you going to turn off the imagery request?" Ramon asked, deflated.

Garabedian thought for a moment. "No, I think *you* should do it. Tell them the vehicle's in excellent shape, and that in the future we will better coordinate to ensure that when a request is made it's done through the proper channels."

Ramon found himself shaking when he got back to his car. Gripping the steering wheel, he took a few deep breaths. He could not bring himself to go back to his office. After a while, he picked up his cell phone and made a call.

"Hello, Rachael."

"My goodness Ilan, you sound upset. Are you all right?"

"I've been better. Listen, I know this is very short notice, but could we meet for lunch?"

"Of course, sweetie. How about Phở Hoang in an hour?"

Phở Hoang Restaurant
Houston, Texas
January 21, 2003

She was the Chinese-American girl with the almost Jewish name.

"What a treat to be able to see you for lunch," Rachael Chen said.

Ilan Ramon gave his wife a kiss before sitting down. "Thank you for making the time."

"You sounded upset on the phone," Chen said, "and believe me, it's no trouble. I spend far more time in sales meetings than any engineering director ever should."

The waiter came and wrote their order for two bowls of beef noodle soup. He then gathered up the menus and took Ramon's cane.

"Excuse me, please." Ramon pointed at his cane. "I need that."

"Oh, sorry sir."

Chen shook her head, then said, "So, tell me what happened."

Fifteen months in an Iraqi prison had not been kind to Ilan Ramon. When he was finally released in the fall of 1982, he weighed seventy pounds and had to be carried out on a stretcher. He spent three months at the Chaim Sheba Medical Centre in Ramat Gan before being sent to the Walter Reed Medical Centre in the U.S. for specialized musculoskeletal reconstructive surgery.

It was there that he met Rachael Chen. Her father was a U.S. Army colonel who was at Walter Reed for a prosthetic fitting. She noticed he was alone in his hospital room, and one day, brought him a steamed bun to share.

Chen went to the University of Maryland to study geology and Ramon followed her there, pursuing a degree in electrical and computer engineering. Upon graduation, Chen was hired by a "dinosaur juice company" (her words) in Houston and Ramon followed her again, eventually getting a job with Rockwell. One night, they went out to watch the Schwarzenegger film *Total Recall* and shortly thereafter, completely on a whim, Ramon submitted his résumé to NASA. He was hired less than a year later.

"I don't believe this," Chen said. "Garabedian expects *you* to do her dirty work for her?"

"I guess so."

"Do you have time for this?" Chen asked.

"Time for what?"

"Well," Chen said slowly, "you're a very busy person, you know. You might not have time to get to this until later today, maybe even tomorrow."

Ramon stared across the table at the beautiful Chinese-American woman with the almost Jewish name. "I love you," he whispered.

NASA Johnson Space Centre, Building 4
January 22, 2003

Ramon had just gotten up from his desk when the phone rang. He was about to let it go to voicemail when he noticed the Colorado area code on the call display.

"*Is this Mr. Ilan Ramon?*" asked a male voice.

"Yes, this is he."

"*Good afternoon, sir. This is Lieutenant-Colonel Timothy Lee at Peterson Air Force Base. I'm calling in regard to your email cancelling the imaging request.*"

"Yes," Ramon said.

"*Well sir, I'm afraid this is a bit awkward, but – the images have already been obtained.*"

Ramon gripped the handset tighter.

"*What would you like us to do with them, sir? Do you still want them?*"

"Yes," Ramon said immediately. "Absolutely, please."

"*Understood, sir. I do need to remind you the images are classified, so they can only be transferred to and viewed by a member of your team with Top Secret clearance.*"

"That would be our flight dynamics officer, Richard Jones," Ramon said.

"*Very good, sir. Please contact Mr. Jones, and I can set up the secure file transfer within the hour.*"

Mission Evaluation Room
NASA Johnson Space Centre, Building 30
January 22, 2003

There was a ghostly beauty to the greyscale images of *Columba* in orbit. The white and grey delta-shaped Shuttle stood out starkly against the black of space. *Columbia* looked almost serene and, in the words of Maria Garabedian, appeared to be in excellent shape.

Except for the hole its left wing.

"Of the dozen or so images obtained, these three provide the best views," explained Richard Jones. "The portside payload bay door obscured the other views." He used a laser pointer to indicate the dark square on the leading edge of *Columbia*'s left wing. "This is *not* a shadow or an artefact or a glitch. It appears in both of the visible light images as well as the near infrared image, in different lighting conditions and different viewing angles."

Ilan Ramon approached the screen and touched the black spot with his finger, as if it were a smudge he could wipe away. "How big is this?"

"Approximately six inches square," said Jones.

"I'm not sure I see how such a conclusion can be made from these images," Peter Delaney said.

Jones turned to Delaney. "The pictures I'm showing you have been down-sampled and blurred to protect the true capabilities of the imaging assets. The Air Force assessment was based on the classified full-resolution images, which I have also seen."

Ramon's cell phone rang. He stepped to the back of the room to take the call.

"*Mr. Ramon? This is Lieutenant-Colonel Lee again at Peterson. I just left a message on your office phone, but I hope you don't mind me calling your cell.*"

Ramon listened. "Lieutenant-Colonel Lee, I need to ask a favour," he said at last, walking back to the conference room table. "Please call me back at this number right away."

A moment later, the phone on the table rang.

"Lieutenant-Colonel Lee," Ramon said, "I have you on speaker phone in the Mission Evaluation Room with my colleagues. Please repeat for them what you just told me."

"*Well sir, it's a little unusual to –*"

"Lieutenant-Colonel Lee," Ramon said, "my colleagues and I believe we are dealing with a safety-of-flight issue aboard *Columbia*. Any information you are able to provide in an open forum would be extremely helpful."

There was a pause, and then the Air Force officer spoke. "*Following the execution of your imaging request, our team went back through tapes of radar tracking data, and we found something. On Flight Day 2, the day after launch, sometime between 15:30 UTC and 16:00 UTC, radar detected a small object, approximately six inches in size, drifting away from the Shuttle orbiter.*"

"Jesus Christ," Rocha muttered.

Ramon pointed at the screen. "I want this analysis documented, and I want a tiger team stood up to assess the implications for re-entry. Thermal, structural, mechanical, guidance, every relevant engineering discipline. Call anyone you need, any building, any NASA centre, any contractor. We present to the Mission Management Team first thing tomorrow morning. And technical reports, *please*. Not just PowerPoint!"

Mission Evaluation Room
8:00am CST, January 23, 2003

It was standing-room only in the drab, grey conference room. In attendance representing management were Maria Garabedian, MER manager

Don McCormack, and Space Shuttle programme manager Ron Dittemore. Jefferson Howell, the director of the Johnson Space Centre, Bryan O'Connor, the chief of safety and mission assurance at NASA Headquarters, and Wayne Hale from the Kennedy Space Centre were tied-in via teleconference. This time, Ilan Ramon and Joyce Seriale-Grush sat near the front.

"The timeline of the launch day in-flight anomaly was compiled by Joyce's team with the support of Richard Jones and Rodney Rocha, and is based on four lines of evidence," Don McCormack reported. "The film from ground cameras at the Kennedy Space Centre, visual and infrared imagery from Air Force assets, radar tracking data also from Air Force assets, and off-line engineering analyses and computer simulations.

"At T-plus 81.7 seconds after launch, when the vehicle was at an altitude of 65,600 feet and travelling at Mach 2.46, a piece of insulating foam with a volume of 1,360 cubic inches separated from the negative-Y bipod ramp area of the external tank, and at T-plus 81.9 seconds struck the lower half of reinforced carbon-carbon panel eight on the leading edge of *Columbia's* left wing. From the relative impact velocity of 840 feet per second, as well as the size and density of the foam and the angle of impact, the imparted kinetic energy was sufficient to breach the RCC panel, resulting in a six-inch hole in the leading edge."

Audible gasps and whispers of "My God" and "oh, shit" drifted through the room.

"Are there any questions or comments at this point?" McCormack asked.

"The USAF imagery was obtained through informal channels," Maria Garabedian said. "We'll need to be more stringent about the JSC-NIMA interface in the future."

"Yes, thank you," Ron Dittemore said. "What are the implications for re-entry?"

"I'll turn it over to Joyce," McCormack said.

"The implication for re-entry is that it's not happening," said Joyce Seriale-Grush. "The thermal team estimates plasma will burn-through the wing leading edge spar between EI-plus 450 seconds and EI-plus 970 seconds, resulting in catastrophic structural failure of the left wing, associated loss of flight control..." Her voice waivered. "And loss of vehicle and crew."

EI was "entry interface", the point about 80 miles altitude when the Shuttle started encountering the effects of the atmosphere. The implication was obvious. *Columbia* would die long before it got anywhere near the safety of a runway.

Ramat HaSharon, Israel

6:26pm IST, January 23, 2003

TV was usually forbidden at meal times, but Omer Bitten made an exception for the duration of his wife's mission. As he set the dinner table, words scrolling on a ticker at the bottom of the screen caught his attention.

"Noa, can you turn that up, please?"

"*…and we have unconfirmed reports of some kind of problem on the Space Shuttle* Columbia. *Unnamed sources are telling us the problem is very serious and may even cut the mission short.*"

The phone rang. Noa picked up, then passed the handset. "It's for you, daddy."

"Who is it?" he asked.

"It's Mr. Aby Har-Even."

Mission Evaluation Room
NASA Johnson Space Centre, Building 30
2:27pm CST, January 23, 2003

"The first priority is conservation," Ilan Ramon said. "We need to conserve every resource, every consumable, in order to buy ourselves time to figure out what to do. So, let's start with power."

"My team is working on a power-down procedure," said Max Leistung, the lead power engineer. "Our goal is to bring peak power down to 9.5 kilowatts, maybe less."

"OK. Food and water?"

"I actually have good news for you, Ilan," said Katie Rogers, the operations lead for environmental controls and consumables. "As long as the crew restricts their physical activities, we've got enough food up there to last more than a month. Assuming the minimal power level, we should still get about three gallons of potable water per crewmember per day as a by-product of the fuel cells."

"That's good news, Katie. Thank you. What about oxygen?"

"We're good for oxygen for thirty-one days," Rogers said. "Actually, the most serious limitation is not too little of something, but too much."

"You mean carbon dioxide," said Joyce Seriale-Grush.

Rogers nodded. "There are sixty-nine cans of LiOH aboard *Columbia*." She pronounced it "lye-oh", referring to the lithium hydroxide canisters used to scrub carbon dioxide from the air. "To estimate how long we could stretch those, I need more information about crew metabolic rates and what is the highest acceptable CO_2 concentration."

"Who would be the best person to ask?"

"Jon Clark."

The room fell silent. Jon was the chief of the Medical Operations Branch at JSC. He was also the husband of *Columbia* astronaut Laurel Clark.

"We can't talk to Jon now," said Peter Delaney. "He's an emotional wreck."

"I don't agree," Seriale-Grush said. "The worst thing we can do is *not* talk to him. It's not as if we're going to hurt his feelings by accidentally reminding him that his wife is stranded in orbit on a broken spaceship. He'll want to help."

"In that case, I'll go talk to him," Delaney offered.

"No, don't trouble yourself," Ramon said quickly. "I'll do it. In the meantime, Max, we need your team to finish that power-down procedure as soon as possible."

Clear Lake, Texas
4:03pm CST, January 23, 2003

Ilan Ramon navigated through an obstacle course of media vans before pulling into the driveway of the Clark residence. Jonathan and his eight-year-old son Iain greeted him.

"Hello, Jon." Ramon embraced him. Iain's arms were wrapped around his father's waist.

"Please, come in," Clark said.

They sat at the dining room table. The window curtains were drawn.

"Jon," Ramon said, "I need to pick your brain. We've got sixty-nine LiOH cans on board, and we need to know how long we can stretch them. How much CO_2 can the crew tolerate over an extended period of time?"

Clark thought for a moment.

"Well, the first thing is to reduce the metabolic rate of the crew. Let's say that we've got them on a 12-hour awake and 12-hour asleep cycle. We assume absolutely minimal physical activities. The currently accepted CO_2 partial pressure for Shuttle and Station is six millimetres of mercury. Mission rules require flight termination if CO_2 goes above 15 millimetres. I've seen papers saying that levels as high as 26 millimetres might be tolerable without long-term effects. So, 15 millimetres should be all right, but it certainly won't be a pleasant experience."

"What are the symptoms?" Ramon asked.

"Shortness of breath, fatigue, headaches. Likely cognitive impairment."

Ramon opened his laptop and brought up a spreadsheet provided by Katie Rogers. "All right, we assume crew metabolism consistent with 12-hour sleep and wake cycles and maximum CO_2 of 15 millimetres, with an assumed absorption capacity per LiOH can, which drives the canister change-out schedule..." He created a chart, then turned the screen to Clark.

"Get Katie to check that, but it looks reasonable to me."

"We're nearing the end of Flight Day 8. This tells us we have until Flight Day 30 before CO_2 levels become unacceptable. That takes us to Friday, February 14th."

"Valentine's Day," said Clark, his voice breaking.

Ramon reached across the table and took Clark's hand. "Then we'll have Laurel home for Valentine's Day. We'll have them all home."

Clark managed a weak smile. "You sound pretty confident."

"I am," Ramon said, "because you're helping."

Aboard Columbia
Flight Day 8
23:18 UTC, January 23, 2003

Ben Hernandez, the commander of *Columbia*, distributed printouts of the power-down procedures to the crew.

"Willie, K.C., Mike, and I will be responsible for power-down of the flight deck and middeck," he said. "First, the general purpose computers. We leave GPC1 on for vehicle control and GPC3 running at 25% for systems monitoring. GPC5 will be in sleep mode and the remaining GPCs will be turned off. We'll stow the Ku-band antenna, deactivate fuel cell three, and power down cooling loop two.

"K.C. and Mike, I need the two of you in the middeck to deactivate the avionics bay instrumentation, set the cabin air fan to low speed, and power-down the galley."

"We're on it," said the flight engineer, Kalpana Chawla.

"Yael, Laurel and Dave, the three of you will be responsible for powering down the SPACEHAB. All payloads and related equipment are to be powered off, including all cameras, TV monitors, and video equipment."

"Understood," said Yael Dahan.

"And Yael, the lab animals in the SPACEHAB...they'll need to be euthanized."

Dahan pursed her lips and nodded grimly.

NASA Johnson Space Centre, Building 1
7:00am CST, January 24, 2003

"How do we to bring them home?" Don McCormack asked. "What options do we have?"

"Can they make it to ISS?" asked Maria Garabedian.

"Impossible." Richard Jones shook his head. "*Columbia* and Space Station are in different orbital inclinations. Russian Soyuz won't work for the same reason."

"So, we'll have to go up and get them," McCormack said.

"Yes, and for that we might just have a chance," Ilan Ramon said. "Angela?"

Angela Brewer, the processing flow director for the Shuttle *Atlantis*, was on the phone from the Kennedy Space Centre. *Columbia's* sister ship, she explained, was already in the Orbiter Processing Facility undergoing preparations for a launch on March 1st. "Atlantis *is in OPF-1 with its three main engines already installed, but no cargo yet in the payload bay. The solid rocket boosters are already attached to the external tank in the VAB,*" Brewer continued, referring to the Vehicle Assembly Building. "*By working three round-the-clock shifts seven days a week, expediting vehicle checks and consolidating or skipping less critical tests,* Atlantis *could be ready for launch by February 9th.*"

"This sounds promising," said Don McCormack, "but it's predicated on two big assumptions. First, that in the expedited launch processing we make no mistakes and don't break anything on *Atlantis*, and second – maybe more importantly – that we are willing to expose *Atlantis* to the same risk of debris strike that's crippled *Columbia*."

Peter Delaney spoke up. "I think it's a mistake, launching again without fully understanding what happened the last time. Do we really want to throw another crew up there?"

"I can guarantee you," Joyce Seriale-Grush said slowly, "that we're going to have astronauts lined up around the block to volunteer for this mission."

Ramat HaSharon, Israel
6:35pm IST, January 28, 2003

Omer Bitten and Noa Dahan sat at the dining table, staring at the television.

"*It is a solemn day at NASA as the space agency recognizes the anniversaries of the* Challenger *explosion in 1986 and the Apollo 1 launch pad fire in 1967, while struggling with the current crisis of seven astronauts stranded in space aboard the damaged* Columbia.

"*In the midst of the darkness, there was a welcome ray of hope as NASA approves a daring rescue mission. The Space Shuttle* Atlantis *was rolled out of the*

Vehicle Assembly Building this morning and began its slow, six-and-a-half hour journey to the launch pad." Under a "recorded earlier" caption, the Shuttle orbiter stacked with its external tank and twin solid rocket boosters rode atop the massive grey crawler vehicle, moving at a stately two miles per hour. "*NASA will complete launch preparations at the pad, working towards lift-off on Sunday, February 9th on a mission to rescue the crew of* Columbia."

"Noa, would you like some food?" Omer asked. She shook her head.

"*We have a video from NASA that shows how the rescue will work.* Atlantis *will rendezvous with* Columbia *about a day after launch.*" On the animation, *Columbia* was upside down with its payload bay towards the Earth. *Atlantis* approached from behind and below *Columbia*, payload bay facing upwards, the vehicle oriented perpendicular to its sister ship. "*The ninety-degree orientation will enable the two Shuttles to get as close together as possible, less than twenty feet apart, without their tails hitting each other.*"

"*Once in position, two spacewalking astronauts from* Atlantis *will connect a tether to* Columbia *to facilitate transfers between the ships. They will first provision* Columbia *with extra spacesuits and carbon dioxide scrubbers, and then they will begin transferring crewmembers from* Columbia *to* Atlantis. *The process is expected to take up to nine hours.*"

Omer squeezed Noa's hand. "You need to eat. Let me get you some food."

Aboard Columbia
Flight Day 17
February 1, 2003

There were some in the ultra-Orthodox community, probably the same ones that had issues with Yael Dahan's gender, who also complained that she would be working on Shabbat while in space. She would be pleased to tell the Haredi this was no longer a problem.

After a brief flurry of activity to power-down the vehicle almost ten days ago, there was now little for the crew of *Columbia* to do except wait and try not to move or breathe too much. At the start of every 12-hour cycle, Dahan emerged from her bunk to use the facilities, eat a cold food pack, and drink her water ration. And that was it. There was literally nothing else to do for the next eleven-and-a-half hours except to eat and drink, and finally to sleep again.

Dahan and her crewmates spent hours staring out the windows at the Earth or the blackness of space. But they couldn't watch movies or listen to music because their battery-powered electronics were dead, and those needing plug-in power had been turned off and stowed. Emails could not be sent or received for the same reason.

She missed Noa and Omer terribly.

History would record that the first Israeli astronaut spent most of her mission floating in her bunk, unintentionally observing Shabbat.

Mission Control Houston
NASA Johnson Space Centre, Building 30
8:52pm CST, February 9, 2003

"*This is Shuttle Launch Control, we are at T-minus 9 minutes and holding. We are awaiting the go/no-go poll of the launch controllers to come out of the hold.*"

Ilan Ramon watched Steve Stich's flight control team in action from the glassed-in VIP area behind and above the Mission Control room. The screen showed an image of *Atlantis* on the pad at Kennedy, glowing in the darkness under the illumination of spotlights.

"Hi, Ilan." Maria Garabedian came into the VIP box, shuffling across the row of chairs. She pointed to the spot beside Ramon. "Is this seat taken?"

"Yes. Joyce will be joining us shortly."

Garabedian moved Ramon's cane aside and sat down anyway.

"*Houston, Launch Control. We are 'no-go' on Range due to an unauthorized aircraft in the launch danger area.*"

Joyce Seriale-Grush came in. She looked at Garabedian for a moment, then sat down beside her. "What's going on?"

"Still holding, due an unauthorized aircraft in the range," Ramon said.

"I hope that guy has an engine failure," Garabedian muttered.

Ramat HaSharon, Israel
4:57am IST, February 10, 2003

The extended family gathered in the predawn darkness to watch the launch. Noa Dahan served tea to her grandparents Michal and Eitan. Omer Bitten chatted quietly with his sister Neta and their parents Rivka and Moshe. The matriarch of the family, Noa's 87-year-old great-grandmother Anat, was dozing in an armchair.

On the television, the illuminated numbers on the countdown clock at the Kennedy Space Centre were changing again. Omer turned up the volume.

"...*after security forces escorted an unauthorized aircraft out of the launch area.*" The picture changed to a view of the launch pad. "*We are now less than seven minutes away from the launch of* Atlantis."

Mission Control Houston
NASA Johnson Space Centre, Building 30
9:01pm CST, February 9, 2003

"Atlantis, *you are 'go' to close and lock your visors.*"

"*Roger, Launch Control,*" said *Atlantis* commander Mike Bloomfield. "*And please tell the crew of* Columbia...*tell them, we're coming.*"

The screen showed a view of *Atlantis* on the launch pad, steam rising and lights glistening in the darkness.

"*T-minus 31 seconds. We have a 'go' for auto sequence start. The five computers on* Atlantis *are now in control.*"

The screen switched to a close-up shot of the engine nozzles at the base of *Atlantis* beneath its tail.

"...*eight, seven, we have a 'go' for main engine start...*"

Orange-red flames spewed as the engines roared to life, quickly turning a translucent bluish-white as the combustion of the liquid hydrogen fuel became complete.

"...*three, two...*"

The flame suddenly disappeared.

"...*and...we have main engine cut-off.*"

Ramon's eyes widened. Joyce Seriale-Grush and Maria Garabedian gasped.

"*We have an RSLS abort. GLS safing and APU shutdown are in progress.*"

Ramon struggled to follow the acronym-filled chatter between Mission Control, the Launch Control Centre, and the crew aboard the Shuttle. There were references to the ground launch sequencer and the redundant set launch sequencer and the auxiliary power units. The screen switched to a close-up of *Atlantis*. He saw water being sprayed on the engine nozzles.

"*Flight, Booster.*"

"*Go ahead, Booster,*" said Steve Stich.

"*Flight, I'm getting a report from the LCC that the MPS fire detectors have tripped.*"

"*Copy, Booster,*" Stich said with deliberate calm. "*Flight controllers, listen up. We are in a Mode One Egress situation.*"

Seriale-Grush put a hand to her mouth.

"The main propulsion system fire detectors went off," Ramon said grimly. "They are evacuating the crew."

Ramat HaSharon, Israel
5:36am IST, February 10, 2003

The family shouted and pointed at the television. Omer Bitten waved his arms and asked for quiet.

"...*hearing now from NASA that* Atlantis *suffered a major engine malfunction less than two seconds from launch. The astronauts have been evacuated and are reported to be uninjured and safe.*"

"Oh, the *fuss*," cackled 87-year-old Anat Dahan. "Why can't they put more gas in that thing and *go* already!"

The illuminated countdown clock was on TV. All the digits were zero.

Mission Evaluation Room
NASA Johnson Space Centre, Building 30
11:23pm CST, February 9, 2003

"What the fuck just happened?" Maria Garabedian shouted.

"The preliminary assessment from Marshall and Rocketdyne indicates a catastrophic failure of the high-pressure liquid hydrogen turbopump in the number two engine," Steve Stich said, "triggering the RSLS abort at T-minus 1.9 seconds." He spread his hands. "I can't...I can't even describe how bad this could've been. If this'd happened two or three seconds later, after we lit the solid boosters, we could have lost the vehicle, the pad, the crew...everything."

"How long to replace SSME-2?" Garabedian asked.

"*We need to change-out all three engines,*" said Angela Brewer on the speaker phone from the Kennedy Space Centre, "*because they all fired and the subsequent water damage.*"

"How long?"

"*Three weeks. Minimum.*"

"The crew will be dead in five days," Garabedian said.

"Yes, thank you," Rodney Rocha muttered.

Garabedian whirled. "Has it ever occurred to you, Rodney...Has it occurred to anyone, that maybe...maybe it would have been better if we didn't know? The crew would have had sixteen wonderful days up there, a terrific mission, and then on re-entry they just, they just..."

A terrible, oppressive silence gripped the Mission Evaluation Room.

After a while, Ramon spoke in a quiet voice. "We'll have to revisit the repair options. The crew will need to fix the damage somehow, and attempt to come home themselves."

Every pair of eyes turned to Ramon.

"How?" Garabedian asked. "They have no repair materials. They are physically and mentally exhausted." She shook her head. "Ilan, you're asking for the impossible."

Ramon closed his eyes. When he spoke again, it was in such a soft voice that people strained to hear. "I am impossible."

"I'm sorry?"

"I am impossible," Ramon repeated. "My mother was a Holocaust survivor. I was born to a woman who wasn't supposed to be alive. When I was eighteen, I had to jump out of an airplane without an ejection seat. When I was twenty-seven, I was blown out of the sky over Iraq. The first time I came to America, I learned to fly the F-16. The second time I came to America, I learned to walk again. And then I learned other things, too."

After a moment of silence, Joyce Seriale-Grush said quietly, "Are you going to say something about failure not being an option?"

Ramon managed a weak smile. "Failure is always a possibility. But giving up, that is not an option. For my part, I will never give up, and I mean never."

People looked at each other, and nodded. In ones and twos, and then in small groups, the engineers stood. They all got up, and with quiet, dignified resolve, went back to work.

Johnson Space Centre, Building 1
11:30am CST, February 10, 2003

"To paraphrase my colleague Ilan Ramon," Don McCormack said, "I think we have our Mission Impossible."

"Let's hear it," said Ron Dittemore, the Space Shuttle programme manager.

Ramon watched Joyce Seriale-Grush move to the front of the room. She looked exhausted, but pressed ahead.

"Our team has devised a repair scenario that we believe to be logistically feasible, making use of existing materials aboard *Columbia*. However, given the critically short timeline, the physiological condition of the crew, and the high number of uncertainties, our team considers the scenario to be high risk with low probability of success."

Seriale-Grush put up series of screen shots from a computer animation. "The crew will fill a bag with whatever pieces of metal they can scavenge from the cabin – metal tools, small bits of titanium, cutlery…any metal they can find. Two astronauts will conduct an EVA, taking the metal scraps and the middeck ladder to use as a work platform, to the site on the left wing."

She pointed to a rendering of two spacesuited figures at the end of a ladder, hovering over an area of the wing highlighted by a yellow square. "They will stuff the bag of metal into the hole, and then they will put a second water-filled bag over it and secure everything as best they can with kapton tape. We will manoeuvre *Columbia* to orient the left wing towards deep space, cold-soaking the patch and freezing the water. *Columbia* will then attempt a modified re-entry profile, one that minimizes the heating of the left wing, hoping to survive long enough to reach bailout altitude."

There was stunned silence.

"A bag of scrap metal and a bag of ice? That's not going to stop three thousand degree plasma!" Maria Garabedian exclaimed.

"No, it's not," Rodney Rocha said, in a trembling voice that betrayed despair. "All we're trying to do is buy time, delay the burn-through of the wing spar, delay the structural failure of the wing for as long as possible. Maybe – just maybe…long enough for the crew to bail out."

Dittemore stared at the screen. Finally, he asked, "What needs to happen next?"

"Four things," Ramon said. "First, we need a team to come up with a detailed EVA procedure for the crew. Second, we need a team to develop the software load for the modified re-entry profile. Third, we need to interface with the Navy for the rescue and recovery effort."

"Because –"

"The re-entry trajectory will be based on the profile for a landing at Edwards Air Force Base in California," Ramon explained. "After the crew has bailed out, we will need to ditch *Columbia* into the Pacific."

"Because we don't want debris falling over Texas. Understood. What's the fourth thing?"

"Someone will need to brief and to also…um, 'manage' the management, here as well as at Headquarters in Washington."

"Let me take care of the politics." Dittemore pointed at the screen. "I need you to make *that* happen."

Ramat HaSharon, Israel
7:47pm IST, February 10, 2003

Omer Bitten lay in bed. He was awake but unmoving, his arms clutching a pillow.

The phone rang. He heard Noa Dahan calling to him, but did not respond. After a moment, he heard his stepdaughter pick up the phone and say something.

She came into the master bedroom.

"Who was that?" he asked.

"Mr. Har-Even," she replied.

"What did he say?"

When Noa didn't respond, Omer rolled over to face her. "What did Mr. Har-Even say?"

"He said eema is going to fix the Space Shuttle."

Aboard Columbia
Flight Day 27
13:13 UTC, February 11, 2003

For the first time in weeks, Yael Dahan felt truly awake and alert. She and Dave Brown were wearing masks and inhaling pure oxygen. This pre-breathe protocol was standard preparation for spacewalks, needed to purge nitrogen from their bodies to prevent the bends. But for Dahan and Brown, the contingency EVA crewmembers, it was also a welcome respite from the CO_2-induced malaise that had tormented them for the past nineteen days.

Kalpana Chawla and Mike Anderson were taping towels over one end of the ladder. Willie McCool did the same with the boots of the spacesuits. Laurel Clark floated about with a large stowage bag, filling it with whatever bits of metal she could scavenge from the crew cabin.

Dahan watched her crewmates with concern. In contrast to her regained state of alertness, she saw they were clearly lethargic and perhaps even cognitively impaired.

Ben Hernandez and Willie McCool helped Dahan and Brown don their bulky spacesuits and enter the airlock. Anderson and Clark passed them the ladder, the bag of scrap metal, and some empty water and stowage bags. They began to close the hatch.

"Jesus Christ, *stop!*" Brown exclaimed. He held out his hands. "Gloves, people. *Gloves!*"

Dahan stared down at her own bare hands, shocked. Seven sets of eyes, seven people with checklists, seven of the smartest people in the world, had somehow managed to miss something that obvious. It was not an auspicious start to the spacewalk.

Mission Control Houston
NASA Johnson Space Centre, Building 30
8:49am CST, February 11, 2003

"Flight, EVA," said Victor Tsang, the extravehicular activity officer. "They are coming out now."

"Copy, EVA," said Ilan Ramon.

The screen showed a shot from a hand-held camera operated by flight engineer Kalpana Chawla, looking out the aft flight deck windows that had a view into the payload bay. Since the contingency spacesuits worn by Yael Dahan and Dave Brown had no cameras, this was the only way Mission Control could see them, and even this view would disappear when the astronauts went over the side of the payload bay to attempt the wing repair.

A white-suited figure emerged from the airlock.

"*Columbia*, Houston for Yael," CAPCOM Stephanie Wilson said. "Congratulations on becoming the first Israeli to walk in space."

Ramat HaSharon, Israel
4:54pm IST, February 11, 2003

On the television, a white-suited figure emerged from the airlock.

"*...are coming out of* Columbia now. *Yael Dahan is wearing the all-white spacesuit, and her colleague the American astronaut Dave Brown can be differentiated by the red stripes on the legs of his spacesuit.*"

"Look, it's eema!" Noa Dahan pointed.

Omer Bitten nodded.

Aboard Columbia
Flight Day 27
15:08 UTC, February 11, 2003

Yael Dahan and Dave Brown floated from the airlock to a tool storage container located at the forward-left corner of the payload bay. Brown retrieved a portable tool caddy called a mini-workstation and attached it to the front of his spacesuit. Dahan took out a payload retention device, a pair of scissors, and a roll of kapton tape from storage and mounted them to the caddy on Brown's suit.

"All right, I'll get Laurel's bag of goodies." Brown went back to the airlock, returning a few minutes later with the metal scraps. As Brown held the bag open, Dahan threw in the remaining metal tools from the container. They returned to the airlock to fetch the ladder.

"All set," Brown said. "I think we've got everything."

Dahan keyed her radio. "Houston, we are proceeding to the work site."

The two spacewalkers floated to the port-side longeron sill, tethering themselves to a slide wire that ran the length of the payload bay. Each holding an end of the ladder, they used their free hands to pull themselves along a set of handrails. It was a gruelling traverse. The stiffness of their inflated gloves and suits made it a physical effort just to maintain a grip.

Dahan and Brown stopped about two-thirds of the way down the length of the payload bay, and climbed over the portside door. They flipped the ladder over the edge and carefully lowered it until the towel-wrapped end touched the left wing, securing the top rung to a door latch using the payload retention device.

"So, there's our work platform." Brown was breathing heavily. "Let's…let's harvest ourselves some thermal blanket."

"Are you all right?" Dahan asked.

"F-fine. Let's go."

They returned to the longeron handrails and continued to the back of the payload bay. *Columbia*'s tail and bulbous engine pods towered over them. After several frustrating attempts, Dahan finally managed to cut a jagged eight-by-seven inch piece of insulation blanket from the aft bulkhead. They returned to the ladder hanging over the edge of the payload bay.

Dahan climbed down. The grey leading edge of *Columbia*'s left wing filled her vision. Except it was not all grey. There, on the curved bottom of the now infamous RCC panel eight, was a gaping black hole. Her eyes widening as the extent of the damage.

"Houston, I am looking at the RCC panel, and the breach looks bigger than six inches, maybe more like seven at least."

"*Can you be more precise?*" asked Mission Control.

"I don't have a tape measure, Houston," Dahan said, immediately regretting her tone.

Brown was now on the ladder behind her. "Will you look at that?"

"Houston, we are proceeding with the repair. We'll do the best we can."

Brown handed Dahan an empty stowage bag. She gently stuffed it into the hole, leaving the mouth open to the outside. Brown started passing her pieces of metal from the other bag, which she in turn placed carefully into the cavity.

"All right, that's the last of Laurel's goodies," Brown said. "I'll be back with the hose."

Dahan took deep breaths, fighting fatigue and willing herself to stay awake as she waited for Brown to return with the water line. The Earth passed beneath her, and she watched the approaching west coast of Africa with tired apathy.

Brown handed Dahan a water container with the hose already attached. She gently placed the container into the hole over the bag of metal.

"All right, K.C.," Dahan called to flight engineer Kalpana Chawla inside *Columbia*. "You can start the water."

As the container inflated, Dahan pushed and kneaded the bag like a lump of dough, trying to achieve a good fit. The expanding container began to press against the edges of the hole.

"Hey K.C., you can stop now," said Dahan.

There was no response. The bag kept growing.

"Kalpana, *stop*, please," Brown called out in a louder voice. The flow of water halted.

"She's exhausted," Dahan said on a private channel.

Dahan pulled the hose from the water bag. Brown handed her the crude patch of insulation blanket, which she put over the water bag, tucking the sides against the edges of the hole as best she could. Finally, she secured the blanket with strip after strip of yellow-gold kapton tape.

"Houston, I think we're done," she said at last.

"*How does it look?*" asked Mission Control.

Dahan surveyed their work. It actually looked ridiculous, like the parcels overwrapped with packing tape she handled the summer she worked at the Israel Postal Company. She tried to recall the English expression. "Jerry-rigged, Houston. It looks totally jerry-rigged."

"*Understood.*"

"We've done the best we can, Houston. We've done everything we can."

Mission Control Houston
NASA Johnson Space Centre, Building 30
12:47pm CST, February 11, 2003

"They're back inside, Flight," said Victor Tsang.

"Copy, EVA," said Ilan Ramon. "And CAPCOM, please send our thanks and congratulations to the crew for a job well done."

Ramat HaSharon, Israel
8:52pm IST, February 11, 2003

"She did it!" Noa Dahan shouted. "Eema did it!"

Omer Bitten nodded, managing a weak smile.

Mission Evaluation Room

NASA Johnson Space Centre, Building 30
4:02pm CST, February 11, 2003

"Well done, everyone," Ilan Ramon said. "Now, we bring them home. Status, please."

"*Columbia* is in the cold-soak attitude," reported Mike Sarafin, the guidance and navigation lead. "The vehicle is oriented with the port wing towards deep space, away from Earth albedo."

"Thank you," Ramon said. "Richard, where are we with re-entry?"

"Given the need to cold-soak for as long as possible and the Friday deadline for consumables," said Richard Jones, "the only re-entry opportunity is Thursday evening, Houston time. Deorbit burn will take place over the Indian Ocean at 04:05 UTC, that's 10:05pm here in Houston. The vehicle will re-enter 45-degrees nose-up, versus the nominal 40-degree angle of attack. The ditching zone is a 2.5 mile by 1.3 mile ellipse in the Pacific Ocean, centred about 70 miles southwest of Los Angeles. The vehicle is expected to descend to the safe bailout altitude of 34,000 feet about 18 miles short of the ditching zone."

"OK, thanks. What about software?"

"We're using the Ops 3 flight control software from STS-111 as a template," said the software engineer David Deschênes, referring to the last Space Shuttle mission to have landed at Edwards Air Force Base in California. "Changing the reference alpha to 45-degrees is requiring a significant change to the code. We'll have *something* ready for uplink by Thursday, but I don't have to tell you the code will hardly have been verified. There could be a bug in there that could send them to Mars."

"Wouldn't that be something," Joyce Seriale-Grush muttered.

Ramat HaSharon, Israel
4:57pm IST, February 13, 2003

As the Sun went down, the street in front of the Bitten-Dahan home began to fill with media vans. Family, friends, and neighbours gathered for the all-night vigil before the re-entry of *Columbia*. Omer Bitten's sister Neta stayed at his side while his parents Moshe and Rivka prepared snacks in the kitchen. Yael's parents Eitan and Michal were in front of the television, and great-grandmother Anat was, once again, dozing in an armchair.

The doorbell rang, and Omer went to get it. Rabbi Eli Levin from Beyt Knesset Darchei Noam entered and greeted him warmly, then joined the crowd by the TV in the family room.

"...*recovery task force will be led by the* USS Essex *out of San Diego, with support from ships of the U.S. Coast Guard and the Maritime Search and Rescue Unit of the Mexican Navy.*" A woman appeared on the screen. "*The crew will bail out over the Pacific, when* Columbia *is below 34,000 feet altitude. This will occur around 9:04pm local time, or 7:04am in Israel.*"

Aboard Columbia
Flight Day 29
01:07 UTC, February 13, 2003

The crew installed their seats in preparation for deorbit burn and re-entry. On the flight deck, Ben Hernandez and Willie McCool would assume their respective commander and pilot positions, with Yael Dahan and Kalpana Chawla sitting behind them. Below in the middeck would be Dave Brown, Mike Anderson, and Laurel Clark.

Brown and Anderson were installing the escape pole that would deploy after they blew the hatch, curving outwards and downwards away from the vehicle. Its purpose was to prevent the astronauts from—ironically— hitting *Columbia's* left wing as they bailed out. Brown would serve as the jumpmaster, helping his crewmates slide down the pole to presumed safety.

"Houston, *Columbia,*" Hernandez reported. "Power-up is nominal, and the Ops 3 Mod 1 flight control software is loaded."

Ramat Aviv Gimmel, Israel
3:37am IST

Joachim Joseph was asleep in front of the TV.

"*We are waiting for a 'go' from NASA for the so-called deorbit burn. This is the engine firing that will take the Shuttle out of orbit and start its plunge into the atmosphere.*"

Stella gently touched him on the shoulder. He snapped awake.

"Yoya, please come to bed," she said.

"I'm not tired!" Joseph whined like a petulant child.

"Of course you're not." She kissed him on the cheek. "Please come to bed. Don't worry, my sweet. I will not let you miss this."

Mission Control Houston

11:05pm CST

"STS-107 flight controllers, may I have your attention please. I would like to invite anyone who wishes to do so to take a moment...to pray, or reflect, or contemplate – whatever is your belief or custom or practice, for the safe return of our colleagues."

Ilan Ramon recited the Tefilat Haderekh silently to himself, then opened his eyes and scanned the room. Some of his colleagues had their hands clasped in front of them. Other sat in silence with their eyes closed. Some just stared ahead in quiet contemplation.

"All right, everyone," he said at last. "Let's bring them home."

Aboard Columbia

The view out the windows began to change from the inky blackness of space to a diffuse orange-yellow glow. Over the course of a few minutes, the glow steadily brightened to the white hot intensity of a blast furnace. Swirls of fiery plasma streaked across the windows.

But the flight deck was quiet. Except for the occasional perfunctory report from Ben Hernandez to Mission Control, no one said a word.

Yael Dahan thought about the jerry-rigged repair on the left wing. The kapton tape and insulation blanket would have burned away quickly, melting like butter on a frying pan. But every minute, every second the remaining mass of ice and metal delayed the plasma's progression to the wing spar, was another second closer to their bailout altitude.

Dahan closed her eyes and slowly counted to ten. Ah'at, shtayim, shalosh...

She was still alive when she reach ten. So she counted again. And again.

If they could live for ten seconds, they could live for another ten. And another. And another. That was how they were going to make it. Ten seconds at a time.

Mission Control Houston

On the screen, *Columbia*'s re-entry track was plotted on a map as a green curve, arcing from the southern Indian Ocean to the Pacific seaboard of the southwestern United States. The red triangle indicating *Columbia*'s position crept towards the coast of California.

"Altitude 90,000 feet, speed 4,308 miles per hour," reported Richard Jones.

"Flight, MMACS," called Jeff Kling, the mechanical and systems officer.

"Go ahead, MMACS," Ilan Ramon responded.

"Flight, we've just lost four temperature transducers on the left side of the vehicle, two of them on system one and one in each of systems two and three."

"Copy," said Ramon. He turned to Joyce Seriale-Grush. Her face was ashen.

"It's started," she whispered.

Aboard Columbia

"Altitude 43,000 feet, speed 806 miles per hour," Willie McCool called out.

Yael Dahan turned and saw blinking red-and-white aircraft lights out the left-side window, just past Kalpana Chawla's helmet.

"Houston, *Columbia*," Ben Hernandez radioed. "It looks like we have company."

"*Roger*, Columbia," said Stephanie Wilson. "*That would be your buddy Mike Bloomfield. He promised to come for you, and here he is.*"

Mission Control Houston

A new window appeared on the screen, showing the feed from a night-vision camera aboard the T-38 chase plane flown by astronaut Mike Bloomfield. *Columbia* appeared as a ghostly image in shades of green against a black sky with greenish-white speckles of stars.

Audible gasps went through the room. Some of the flight controllers stood from their consoles like an honour guard.

The hole in *Columbia*'s left wing was now an obsidian gash. There were black streaks over the wing and along the fuselage, and dark splatters on the left engine pod and tail—cooled residue of molten metal. The rudder and elevon were deflected, physical manifestation of the flight control system struggling to keep the ship steady.

A chill went down Ilan Ramon's spine. *Columbia* was mortally wounded, but she was still alive, still fighting to bring her crew home. She was simply a beautiful, magnificent, heroic flying machine.

"Don't do it."

Ramon blinked. Had he said something aloud?

"Don't anthropomorphize the vehicle," Joyce Seriale-Grush said. "She doesn't like it."

Aboard Columbia

A buzzing vibration shook *Columbia* as the vehicle passed below the speed of sound.

"Altitude 42,000 feet, and we are now subsonic," Willie McCool said.

"All right, everyone," Ben Hernandez called out. "Final verification of your sea survival gear, and have your breakup/LOC cue card handy."

Yael Dahan finished checking her pressure suit, parachute, and survival gear as McCool reported their altitude at 40,000 feet.

"Houston, *Columbia*," Hernandez said. "We are venting the cabin."

Mission Control Houston

"Cabin venting complete," reported Katie Rogers from the EECOM console.

"Altitude 34,120 feet," said Richard Jones.

"That's close enough," Ilan Ramon said. "CAPCOM, get them out of there."

"*Columbia*," Stephanie Wilson called, "you are 'go' for Mode One Egress."

Aboard Columbia

"Blow the hatch!" Ben Hernandez ordered.

There was a muffled bang as the middeck door jettisoned and the escape pole deployed. Through her helmet, Yael Dahan could hear a faint whistle of wind.

"Jumpmaster is in position," Dave Brown called from the middeck.

"Houston, *Columbia*." Hernandez made the final report. "We are initiating the bailout procedure now."

After disconnecting the oxygen and communications lines from her pressure suit, Dahan unfastened the seatbelt and threw the harness over her head. She followed Kalpana Chawla to the passageway leading to the middeck below. Chawla climbed down the ladder.

Dahan grasped the top rung. She pressed a hand against the bulkhead.

"Toda raba," she whispered, then descended the ladder.

Mission Control Houston

"Flight!"

Ilan Ramon looked up, and watched *Columbia* die.

The left wing simply disintegrated. *Columbia* rolled over violently, pitched down, and then went into a flat spin. The tail snapped off, and then the right wing and the main fuselage came apart, throwing the crew cabin clear.

For a moment, the image disappeared as the debris fell out of the field-of-view of the night-vision camera. Then the camera found the targets again and started tracking downwards, following the pieces of wreckage like giant, greenish-white snowflakes tumbling towards the dark ocean below.

Ramat Aviv Gimmel, Israel
7:20am IST

"No!" Stella cried.

Joachim Joseph pulled his wife closer.

It was the middle of the night in Houston. On the television, a sombre-faced reporter was standing in front of the large NASA sign at the entrance to the Johnson Space Centre. Flowers, wreaths, balloons, and cards were already piling up.

"*We now have a report from NASA, from the flight dynamics officer at Mission Control, that the vehicle is destroyed. We are told that rescue forces will begin moving into the recovery zone as soon as debris has stopped falling.*"

Mission Control Houston
11:21pm CST

"Recovery, how many?" Ilan Ramon asked.

The screen showed a view from a night-vision camera aboard the USS *Essex*. The dark ocean blossomed with a greenish-white plume every time a piece of wreckage hit the water. On an adjacent telemetry screen, there was no more data, only the letter "S" for "static" or "NaN" meaning "not-a-number".

Joyce Seriale-Grush was crying.

Ramon wiped his eyes with the back of his hand. "How many, Recovery?"

"Three, Flight," came the answer at last. "There were three parachutes."

Ramat HaSharon, Israel
9:09am IST

"Three, three, three!" shouted 87-year-old Anat Dahan. "They keep saying three but they won't say who. Who is it?"

The doorbell rang.

Moshe Bitten answered it. "Omer, Noa. I think…I think you better come here."

An icy knot formed in Noa's stomach. She took Omer's hand, and together went to the front door.

It was Aby Har-Even. He was wearing a dark suit.

Noa Dahan covered her mouth and began sobbing.

Omer Bitten let out an anguished, almost inhuman cry of grief and despair. Har-Even and Moshe reached out to steady him as Omer's knees gave and he sank to the floor.

Mission Control Houston
1:10am CST, February 14, 2003

Ilan Ramon sat alone at the Flight Director's console. Mission Control was empty. Screens were dark, consoles were off, and lights were dimmed.

Ramon slowly took off his headset and tossed it to the console. Taking out his cell phone, he sent a text. Not entirely to his surprise, he received a response only a few seconds later. He dialled a voice number.

"You're still up," said Ramon.

"*Of course,*" Rachael Chen said. "*I don't think anyone slept tonight. Ilan, I am so sorry.*"

"Such a terrible day." His voice broke.

"*You and your colleagues did everything you could.*"

"We could have done more. We could have saved more." He knew them to be the words of the son of a Holocaust survivor.

"*Come home, Ilan.*"

"Because I *can* come home."

"*Yes, because you can. So, come home. I'll be waiting.*"

Ilan Ramon took his cane and made his way to the exit. He glanced behind him one last time, then turned out the lights and locked the doors. He was going home.

Lod Air Force Base, Israel
May 12, 2003

It took almost three months for the wreckage of *Columbia* and the remains of her crew to be recovered from the bottom of the Pacific, but at long last, Yael Dahan also came home.

An honour guard carried the casket, draped with the Flag of Israel, down the cargo ramp of the Israeli Air Force C-130 transport. Noa Dahan and

Omer Bitten followed the procession into a hangar where family, friends, colleagues, and dignitaries had gathered.

Noa and Omer took their seats beside Prime Minister Ariel Sharon and former Prime Minister Shimon Peres. Aby Har-Even sat with NASA Administrator Sean O'Keefe. Noa saw the surviving *Columbia* astronauts—Laurel Clark, Kalpana Chawla, and Michael Anderson—and their families. Laurel Clark sat between Jon and Iain, holding their hands.

The casket was placed on a raised platform beside a podium. A large-size image of Yael Dahan's astronaut portrait, and those of Ben Hernandez, Willie McCool, and Dave Brown, flanked the casket. The Chief Rabbi of Israel read "A Woman of Valour" from the Book of Proverbs, and then Hatishma Koli began to play.

"Allow me, on behalf of the people of Israel, to honour the memory of Yael Dahan on this, her last road in our homeland," said Prime Minister Sharon. "Yael's image, projected from above, was a reflection of Israel at its best. As Israel's space pioneer, her memory will be engraved in our hearts forever."

Hand-in-hand, Noa and Omer made their way to the podium. He was wearing a dark suit. She wore her mother's blue NASA jacket.

Omer recited the mourner's kaddish, then started reading from a piece of paper.

"My dear Yael went in search of a better world. In the midst of our sorrow, our family takes comfort in knowing that she died doing what she loved – " he looked at the surviving astronauts—"with people she loved. Everyone who knew Yael knows it is impossible to remember her without a smile on her face, and we will continue forward with that same smile."

He switched to English. "I want to say something now, to the people at NASA. I wish I could talk to every person who worked on this mission, and tell them…" Sobs racked his body, and he collapsed against the podium.

Noa pulled Omer closer and help him stand. Gently, she took the crumpled piece of paper from his hand and smoothed it out on the podium.

"What my father wants to say," she said in a soft but steady voice, "is that our family wishes to thank every person at NASA who worked on this mission, who never gave up trying to bring my mother and her friends home, even when things must have seemed impossible. Because of such people, there will always be goodness and hope in the world."

Noa repeated the words in Hebrew, and then helped Omer off the stage. She looked into his eyes. Behind the tears of grief there was love—and fierce pride.

Department of Geophysics and Planetary Sciences

Tel Aviv University
May 13, 2003

Joachim Joseph picked up the phone. It was Scott Janz, his MEIDEX co-investigator at the NASA Goddard Space Flight Centre.

"Scott, how are you my friend?" Joseph said. "Listen, I know I'm behind on the abstract for the IASTP Conference, but you can be sure –"

"Yoya, I am sorry to interrupt, but I'm not calling about the abstract." Janz paused. "Yoya...they found it."

"Found what?"

"Your Torah scroll. They have recovered it from the wreckage."

A lump formed in Joseph's throat.

"The scroll and ark were sealed in plastic bags and wrapped in some towels and stuffed into a middeck locker. Someone did this deliberately, trying to protect it from the crash. There's some water damage, but it is intact. They found it, Yoya!"

"Oh, Scott..." Joseph took a moment to compose himself. "I don't know what to say. Wonderful. Thank you. Thank you!"

"We're going to get that Torah back to you right away," Janz said. "I promise."

International Atmospheric and Solar-Terrestrial Physics Conference
Binyenei HaUma Convention Centre, Jerusalem
September 2003

"In summary," Joachim Joseph said, "we were successful in determining the height distribution in a set of six areas of desert dust aerosol plumes over the eastern tropical and subtropical Atlantic Ocean, using multispectral reflected radiance data collected by the MEIDEX payload aboard the *Columbia* mission.

"If there are no questions, I will conclude by acknowledging my co-investigators Zev Levin, Yuri Mekler, Peter Israelevich, Eli Ganor, Scott Janz, Ernest Hilsenrath, Mustafa Asfur, Adam Devir, Meir Moalem, and Yoav Yair, as well as all the students in Israel and America who contributed to this project over many years. This work is dedicated to the *Columbia* astronauts, without whom we would not have the data."

As Joseph disconnected his laptop from the projector, a middle-aged man in a light grey suit approached the podium.

"Professor Joachim Joseph?"

"Uh, yes. Just Yoya, please."

"Yoya, hello." They shook hands. "My name is Jeremy Issacharoff. I am from the Ministry of Foreign Affairs."

"Yes," Joseph said. "I remember your email. It is a pleasure to meet you."

"No, truly, the pleasure is mine." The polished-looking diplomat seemed at a loss for words. "As I wrote in my email…well, I know it was a crazy thing to ask, being that you don't really know me at all, but – it would mean so much, if there were any way it could be possible? I only hope you will consider my request."

"Yes, Mr. Issacharoff," said Joachim Joseph. "I have considered your request."

Beyt Knesset Menachem Zion
Jerusalem
November 2003

Four small candles were lit, and then the rabbi called the young man forward.

Thirteen-year-old Dean Issacharoff slowly approached the table and, with white-gloved hands, gently took hold of the small, water-stained Torah scroll. He began to read.

* * *

Jerusalem Post, *Tuesday, December 9, 2008*
Joachim Joseph, Professor of Physics, Tel Aviv University

Joachim Joseph, professor of physics at Tel Aviv University, died Monday at his home in Ramat Aviv Gimmel. He was 77.

Joseph was born on March 31, 1931 in Berlin, Germany. During the Second World War, Joachim (or "Yoya", as he preferred to be called) and his family were sent to Bergen-Belsen. After the war, Yoya and the surviving family members sailed to Palestine to help pioneer the modern state of Israel. Young Yoya excelled in science, eventually earning a doctorate in atmospheric physics from the University of California, Los Angeles before becoming a professor at Tel Aviv University.

Yoya was a world-renowned expert on desert aerosols, becoming the principal investigator for the Mediterranean Israeli Dust Experiment that flew with astronaut Yael Dahan on the ill-fated Columbia *mission in 2003. It was through Dahan that Yoya's death camp experience returned to bookend his life in a remarkable way. Rabbi Simon Dasberg, a fellow prisoner at Bergen-Belsen, secretly trained Yoya for his Bar Mitzvah, afterwards giving him the tiny Torah scroll and extracting a promise from Yoya to tell the story to the world. Almost sixty years later, Yael Dahan took the scroll with her to space.* Columbia *tragically broke apart on re-entry, but the little Torah survived and was recovered from the wreckage. Yoya's family has bequeathed the scroll to the Holocaust History Museum at Yad Vashem. The scroll will be the centrepiece of a new travelling exhibit that will visit more than a dozen countries before returning to its permanent home at the Yad Vashem museum.*

Yoya is survived by his wife Stella, his daughters Iris and Gili, his sons-in-law Ronen and Pablo, and six grandchildren, Yuval, Chen, Liad, Tal, Gal, and Yam. His legacy also continues through his scientific papers and the countless students he mentored over a remarkable life.

Afterword

I was inspired to write "A Sky and a Heaven" by two personal events.

On 1 February 2003, I was at the Kennedy Space Centre to watch the landing of the Space Shuttle *Columbia* at the conclusion of Mission STS-107. My colleagues and I waited for the double sonic boom that announces a Shuttle's arrival, but there was nothing. The landing countdown clock went to zero, and then started counting up. A few minutes later, we were ordered back to our bus. On the ride back to the Operations and Checkout Building, a woman got a mobile phone call from a relative informing her that a TV new channel was reporting that debris had fallen in Texas. *Columbia* had broken apart during re-entry, killing six American astronauts—Rick Husband, Willie McCool, Kalpana Chawla, Laurel Clark, Michael Anderson, and David Brown—as well as the first Israeli astronaut, Ilan Ramon.

Thirteen years later, I was a visiting lecturer at the 2016 summer session of the International Space University (ISU) that was hosted at the Technion Israel Institute of Technology in Haifa. Ilan Ramon's late widow Rona was a speaker on a distinguished panel. I was deeply moved by Ilan Ramon's enduring legacy and the strength with which his memory was being kept alive. Rona Ramon was the founding CEO of the non-profit Ramon Foundation for youth academic excellence and social leadership through science and technology. The documentary filmmaker Michael Potter established the Ilan Ramon Scholarship Fund that provides financial support for Israeli students to attend ISU. There is a street named for Ilan Ramon in a suburb north of my home town of Toronto.

According to the findings of the Columbia Accident Investigation Board (CAIB),[1] at T-plus 81.7 seconds after its launch on 16 January 2003, when the Shuttle was at an altitude of 20 kilometres and travelling at Mach 2.46, a piece of insulating foam separated from the negative-Y bipod ramp area of the external tank (where an inverted V-shaped strut connects the tank to the nose of the Shuttle orbiter), and at T-plus 81.9 seconds struck one of the reinforced carbon-carbon (RCC) panels on the leading edge of the left wing at a relative velocity of 256 metres per second. Beyond the immediate technical causes, the CAIB Report also examined the organizational roots of the accident and was highly critical of what it called "a broken safety culture"[2] at NASA, at one point even lamenting "the endemic use of PowerPoint briefing slides instead of technical papers as an illustration of the problematic methods of technical communication"[3] at the U.S. space agency. The phenomenon of bipod foam shedding was first observed in 1983, but two decades of (mostly) successful missions had lulled management into believing what happened to *Columbia* was only a post-flight maintenance issue rather than an urgent safety-of-flight problem.

As part of its enquiry, the members of the Columbia Accident Investigation Board looked into whether anything could have been done to save the crew. In order to "put the decisions made during the flight of STS-107 into perspective",[4] the Board asked NASA engineers to devise both a rescue option and a repair option. The results of this study were documented in Volume 2, Appendix D.13 of the CAIB Report, which had the rather prosaic title of "STS-107 In-Flight Options Assessment".[5] My research for "A Sky and a Heaven" drew heavily on this document.

The rescue option called for an expedited launch of another Shuttle to rendezvous with *Columbia* and return the crew to Earth. In the repair option, the *Columbia* crew would attempt to fix the wing damage themselves using whatever tools and materials were available to them. Two other options, retreating to the International Space Station (ISS) as a safe haven or using a Russian Soyuz as a rescue vehicle, were impossible because of orbital mechanics. *Columbia* was in a 39 degree inclination orbit (the angle of the orbital plane with respect to the equator), whereas the ISS is in a 51.6 degree

---

[1] CAIB (Columbia Accident Investigation Board). 2003. Report, 6 Vols. U.S. Government Printing Office, Washington, DC. https://history.nasa.gov/columbia/CAIB_reportindex.html.

[2] CAIB Report, Vol. 1, Chaps. 6 and 7.

[3] CAIB Report, Vol. 1, Chap. 7, Pg. 191.

[4] CAIB Report, Vol. 1, Sect. 6.4, Pg. 173.

[5] "STS-107 In-Flight Options Assessment", CAIB Report, Vol. 2, App. D.13, 2003.

orbit and the lowest inclination that Soyuz could achieve is 45.9 degrees due to the latitude of the Baikonur Cosmodrome.

The timeline of events described in Appendix D.13[6] assumed that NASA mission managers requested imagery on Flight Day 3 (18 January) immediately after discovery of the debris strike the previous day. This imagery was assumed to be inconclusive, and combined with the fact that the crew could not see the impact area from the windows, led to a decision on Flight Day 4 to perform an EVA (extravehicular activity, or spacewalk) on Flight Day 5 that confirmed catastrophic damage. In "A Sky and a Heaven", I assumed the imagery request (or rather imagery non-cancellation) took place on Flight Day 5 but there was no need for an inspection EVA because the pictures were sufficient to show the damage on Flight Day 8.

Following confirmation of the damage, the crew would begin conserving resources such as power, oxygen, food, water, and the lithium hydroxide (LiOH) canisters that remove carbon dioxide. By minimizing activities, there would have been sufficient food, water, oxygen, and power to keep the crew alive for more than a month. The constraining resource was the limited number of CO_2 scrubbers, which would have imposed a rescue or return deadline of no later than Flight Day 30 (14 February). Conserving the LiOH canisters would have required allowing the partial pressure of CO_2 to rise from 6 mmHg to 15 mmHg. The detrimental effects of high CO_2 levels—shortness of breath, fatigue, headaches, and possible cognitive impairment—have been documented by the American astronaut Scott Kelly during his almost year-long mission aboard the ISS.[7]

What made the rescue option feasible was an ace-in-the-hole in the form of the Space Shuttle *Atlantis*.[8] The sister ship of *Columbia* was already in the Orbiter Processing Facility at the Kennedy Space Centre in an advanced state of preparation for a planned launch on 1 March. Its three main engines were already installed, the payload bay was still empty, and the solid rocket boosters were already attached to the external tank in the Vehicle Assembly Building. By working three round-the-clock shifts seven days a week, expediting vehicle checks and consolidating or skipping less critical tests, *Atlantis* could have been ready for lift-off as early as 9 February, with backup launch windows over the next two days. According to meteorological records, the weather on

[6] CAIB Report, Vol. 2, App. D.13, Sect. 2.0, Fig. 3, Pg. 398.

[7] Kelly, Scott, *Endurance: A Year in Space, a Lifetime of Discovery*, Penguin Random House, 2017, pp. 86–91.

[8] Hutchinson, Lee, "The Audacious Rescue Plan That Might Have Saved the Space Shuttle Columbia", Ars Technica, Posted 1 February 2016, https://arstechnica.com/science/2016/02/the-audacious-rescue-plan-that-might-have-saved-space-shuttle-columbia/.

those days (both at the Kennedy Space Centre and at the abort landing sites) would have allowed a launch.[9]

The rescue option would have been predicated on two major assumptions. First is that no problems and no mistakes or failures occurred during the rushed launch processing. The second and perhaps more important assumption is that NASA would have been willing to expose *Atlantis* and its crew to the same risk of debris strike that crippled *Columbia*. This would not have been a trivial consideration. In the 112 missions before STS-107, there were six confirmed instances of bipod foam shedding from the external tank (including the previous flight of *Atlantis*, just before STS-107),[10] and overall there were fourteen flights that experienced significant thermal protection system damage from foam strikes and other impacts.[11] In order to make a 9 February launch date and meet the 14 February deadline for consumables, there would not have been time to design and implement any modifications to the external tank.

Assuming a go-ahead, NASA would have launched *Atlantis* with a minimum crew of four, consisting of a commander and pilot to fly the vehicle and two mission specialists to perform the rescue EVA. Due to the rapid schedule of the rescue mission, with no time for in-depth training and no margin for error, the crew would have consisted entirely of veteran astronauts with previous flight experience who were also not susceptible to space adaptation syndrome (motion sickness). In January 2003, there were seven commanders, seven pilots, and nine mission specialists who would have met these requirements.[12]

Assuming a successful launch on 9 February, *Atlantis* would rendezvous with *Columbia* the following day. *Columbia* would be oriented to fly tail-first with its payload bay pointed towards the Earth. *Atlantis* would close-in on *Columbia* from below and behind, its payload bay facing upwards, flying a trajectory called a "positive R-bar approach" (relative to an imaginary radial line between *Columbia* and the centre of the Earth) that minimises thruster plume impingement on its sister ship. Finally, *Atlantis* would come to a relative stop less than six metres from *Columbia*, the two vehicles oriented perpendicular to each other in order to prevent their tails from hitting.

This would have been the first (and presumably only) time that two Space Shuttles were in orbit at the same time, and the operational challenges would

9 CAIB Report, Vol. 1, Sect. 6.4, Pg. 173.

10 CAIB Report, Vol. 1, Fig. 6.1-6, Pg. 127.

11 CAIB Report, Vol. 1, Fig. 6.1-7, Pg. 128.

12 CAIB Report, Vol. 2, App. D.13, Sect. 3.6, Pg. 401.

have been considerable.[13] *Columbia* and *Atlantis* would each have needed its own flight control room at the Mission Control Centre in Houston, operating in parallel with a third flight control room for ongoing ISS operations. According to Appendix D.13, there would have been sufficient ground segment hardware and software capability (and presumably personnel) to support all three.[14]

With *Atlantis* on-station, the two EVA astronauts would leave the airlock to start the rescue operation. With the help of SAFER (Simplified Aid For EVA Rescue) jet packs, the astronauts would either extend a boom or connect a tether to *Columbia* in order to facilitate transfers between the ships. They would provision *Columbia* with extra LiOH canisters and two additional spacesuits before starting the process of bringing crewmembers back to *Atlantis*.

This is much more difficult than it sounds. Donning a spacesuit is a complicated process that normally requires assistance from other crewmembers. For the hypothetical rescue, the task would have been made harder by the fact that the number of available helpers aboard *Columbia* diminished with each successful crew transfer.[15] While this was happening, *Atlantis* would be constantly drifting away from *Columbia* due to Kepler's laws of orbital motion (*Atlantis* being in an ever so slightly lower, faster orbit). The commander and pilot of *Atlantis* would have the exhausting task of manually station-keeping for the entire duration of the operation, which Appendix D.13 assumes would take about nine hours[16] but others have estimated could take up to three times longer.[17]

Prior to leaving *Columbia*, the last two crewmembers would have one final task: to prepare for the scuttling of their ship. Under the rescue option, no attempt would have been made to recover *Columbia* because American space shuttles (unlike the Russian shuttle *Buran*[18]) were not originally designed to land automatically. Mission Control did not have the capability to remotely start the auxiliary power units, deploy the air data probes, or extend the landing gear (cynics have speculated this was done deliberately to ensure the need for astronauts). Before leaving *Columbia*, the last crewmembers would have gone to the flight deck and toggled a number of switches to arm the orbital manoeuvring system engines and give Mission Control remote access

---

[13] Hutchinson, Ars Technica.

[14] CAIB Report, Vol. 2, App. D.13, Sect. 3.5, Pg. 401.

[15] Hutchinson, Ars Technica.

[16] CAIB Report, Vol. 2, App. D.13, Sect. 3.10, Pg. 403.

[17] Hutchinson, Ars Technica.

[18] Johnson, Nicholas L., *The Soviet Year in Space 1988*, Teledyne Brown Engineering, 1989, Pg. 110.

to the five general purpose computers in preparation for an eventual ground command to deorbit the vehicle into the South Pacific.[19]

A final challenge of the rescue option would be bringing the combined contingent of eleven astronauts back to Earth. There would not be enough seating on *Atlantis*, and it is likely that four of the astronauts would literally have to be strapped to the floor of the middeck.[20] Fortunately, Appendix D.13 concluded that the additional weight and changed centre-of-mass would not have materially affected vehicle dynamics during re-entry, and no ballasting or additional propellant would have been required.[21]

The notional timeline in Appendix D.13 called for NASA to pursue both the rescue and repair options until around Flight Day 26 (10 February), after which they would commit to one course of action and abandon the other.[22] In "A Sky and a Heaven", I took dramatic license in portraying sequential attempts at rescue and then repair, although it was hinted that repair options had been considered earlier. Another deviation is that Appendix D.13 called for two EVAs, the first to retrieve tools from the payload bay stowage assembly and harvest a piece of thermal blanket from a less critical area for use in the repair, and the second to attempt the actual repair itself. For the purposes of the story, I combined both into a single spacewalk.

A crucial factor in the feasibility of the repair option was the size of the hole in the wing. Appendix D.13 assumed a relatively modest 15 centimetre (six inch) hole, an assumption that I retained for the purposes of the story. We will never know the actual size of the hole that doomed *Columbia*, but it is likely to have been much larger. NASA estimated that a hole of at least 25 centimetres would have been needed for the wing to ingest a plasma stream large enough to cause the damage observed in recovered debris.[23] A test conducted by the Southwest Research Institute, in which a nitrogen gas gun was used to fire a piece of external tank foam at an RCC panel taken from the Shuttle *Atlantis*, produced a hole of about 41 by 43 centimetres that was big enough for people to put their heads through.[24]

Every Space Shuttle mission carried two spacesuits and a basic set of tools, and two crewmembers were always trained to perform contingency EVA

---

[19] CAIB Report, Vol. 2, App. D.13, Sect. 3.12, Pg. 404.

[20] Hutchinson, Ars Technica.

[21] CAIB Report, Vol. 2, App. D.13, Sect. 3.11, Pg. 404.

[22] CAIB Report, Vol. 1, Sect. 6.4, Pg. 173.

[23] Leinbach, Michael D. and Ward, Jonathan H., *Bringing Columbia Home: The Untold Story of a Lost Space Shuttle and Her Crew*, Arcade Publishing, 2018, Pg. 246.

[24] Langewiesche, William, "Columbia's Last Flight", *The Atlantic*, Nov. 2003, Pg. 77.

tasks like manually closing and latching the payload bay doors. For the STS-107 mission of *Columbia*, the contingency EVA crewmembers were Michael Anderson and David Brown.

To execute the repair option, Anderson and Brown would have gone outside and attempted to fill the hole in the wing with pieces of metal scavenged from the crew cabin. They would then cover the hole with a water-filled bag, which when frozen would hold the scrap metal in place and partially restore the geometry of the wing leading edge. It may seem incredible to think that a bag of ice and some pieces of metal would offer any protection from the searing heat of re-entry, but the hope was that even crudely restoring the shape of the wing might delay the tripping of the boundary layer (the thin region of flow adjacent to a surface where viscous effects dominate) into turbulent flow, a phenomenon that would have significantly increased the thermal loads.

As part of my research for "A Sky and a Heaven", I contacted former NASA flight director John Shannon who led the Appendix D.13 study, and I asked him what metal objects from the orbiter cabin or the SPACEHAB research module would have been available for the crew to use in the repair. He replied that his team only had two weeks to do the study and were not able to get into specifics.[25] I decided to leave this vague in the story. Another unanswered question was whether the plastic bag (in NASA-speak, a contingency water container or CWC) was sufficient to hold the vapour pressure for the water to stay liquid. For the purposes of the story, I took dramatic license and assumed it was.

To freeze the water, the crew would have oriented *Columbia* to point the left wing towards deep space, away from Earth albedo (sunlight reflected off the planet) to "cold-soak" for three to six days. In the story, only two days are available for cold-soak due to the time taken in pursuing the rescue option first. Regardless of duration, this type of cold-soak would never have been attempted on a normal mission due to the adverse thermal effects on systems like the payload bay doors and the landing gear tires. In an early draft of "A Sky and a Heaven", the crew was unable to close the payload doors after the cold-soak because of thermal distortion.

Appendix D.13 considered the possibility of jettisoning some or all of the hardware in the payload bay to reduce the vehicle mass for re-entry.[26] For the STS-107 mission, *Columbia* was carrying a SPACEHAB pressurized research module (about 5,000 kg), a collection of externally mounted experiments called FREESTAR (about 2,000 kg), and an Extended Duration

---

[25] Shannon, John. Email correspondence with the author, 18 Dec. 2017.
[26] CAIB Report, Vol. 2, App. D.13, Sect. 4.7, Pg. 407.

Orbiter pallet (about 3,200 kg) that carried additional liquid hydrogen and liquid oxygen reactant for the electricity-generating fuel cells. Jettisoning this hardware would have necessitated more EVAs attempting difficult tasks like undoing bolts, cutting power and fluid lines, and handling large free-floating masses that *Columbia* would later have to carefully manoeuvre away from. I disregarded this option in the story because I thought even the one repair EVA would have taxed the astronauts to the limit.

For re-entry, Appendix D.13 suggested raising *Columbia*'s angle of attack to 45 degrees from the normal 40 degrees to reduce the heating on the wing leading edges.[27] A more detailed thermal analysis would have been required to make sure this did not dangerously increase the heating on another part of the vehicle. The higher angle of attack would have required significant changes to the flight software, the development and verification of which would have been extremely challenging in such a short time.

Appendix D.13 did not elaborate on the rest of the re-entry and crew bail-out procedure, so for the story I was left to my meandering imagination. I postulated a re-entry trajectory similar to that for a landing at Edwards Air Force Base in California in order to facilitate a crew bail-out and subsequent vehicle ditching over the Pacific Ocean, and I used STK (Systems Tool Kit) software to estimate a probable timeline of events. The crew would have needed to wait until *Columbia* descended below 10.4 kilometres (34,000 feet) before attempting a bail-out, blowing the middeck hatch and sliding out a four metre escape pole that was installed on the Shuttle fleet after the *Challenger* accident in 1986.

NASA considered the rescue option to be "challenging but feasible",[28] whereas for the repair option the "overall assessment of the expectation of task success is moderate to low, depending on damage site characteristics"[29] with the troubling addendum that "the risk of doing additional damage to the orbiter is high (i.e. enlarging the wing leading edge breach)."[30] According to John Shannon, "The only real option, which also had significant risk, would have been to launch *Atlantis* on a rescue mission."[31] Some of his former colleagues do not believe any sort of rescue would have been possible.[32] In this context, the conclusion of "A Sky and a Heaven" is absolutely the most

---

[27] Ibid.
[28] CAIB Report, Vol. 1, Sect. 6.4, Pg. 174.
[29] CAIB Report, Vol. 2, App. D.13, Pg. 410.
[30] CAIB Report, Vol. 2, App. D.13, Sect. 4.5, Pg. 406.
[31] Shannon, 2017.
[32] Leinbach and Ward, Pg. 269.

optimistic possible outcome of the repair scenario. That is a rather sobering thought.

As you can tell from my previous stories "The Son of Heaven", "The Coming Age of the Jet" and "Divisions", I very much enjoy exploring science-themed alternate history. It is, however, a challenging genre to write due to the sensitivity of putting real people into fictional situations. Authors of alternate history have an obligation to be careful in their portrayals of real people and ensure their words and actions are consistent with what is known about them from the historical record. I changed the name of the commander of *Columbia* to Ben Hernandez because in an early draft of the story this person did something that I felt was inconsistent with the personality of the real commander Rick Husband. It is implied that Ben Hernandez was an ancestor of Erika Hernandez, the captain of the starship NX-02 *Columbia* in the science fiction TV series *Star Trek: Enterprise*.

The story of Joachim Joseph and the little Torah scroll that went from Bergen-Belsen to outer space was told in Daniel Cohen's documentary *An Article of Hope*. This scroll was lost in the *Columbia* accident. Three years later in September 2006, at the behest of Rona Ramon the Canadian astronaut Steve MacLean carried with him aboard the Space Shuttle *Atlantis* a "sister Torah" that belonged to Henry Fenichel, a physics professor at the University of Cincinnati who like Joseph was a survivor of Bergen-Belsen. Following his return from the mission, MacLean returned the Torah to Fenichel in a ceremony at the University of Cincinnati in which seventy Holocaust survivors were in attendance. A separate "return of the Torah" ceremony was held at the Weizmann Institute of Science in Israel where Fenichel met Rafael Ben Zeev, the grandson of Rabbi Simon Dasberg who had given Joachim Joseph his scroll after the secret Bar Mitzvah at Bergen-Belsen in 1944.

CPSIA information can be obtained
at www.ICGtesting.com
Printed in the USA
LVHW022108280822
727031LV00004B/185